Pragmatic Electrical Engineering: Systems and Instruments

Pragmatic Electrical Engineering: Systems and Instruments
William Eccles

ISBN: 978-3-031-79836-8   paperback
ISBN: 978-3-031-79837-5   ebook

DOI 10.1007/978-3-031-79837-5

A Publication in the Springer series
SYNTHESIS LECTURES ON DIGITAL CIRCUITS AND SYSTEMS

Lecture #34
Series Editor: Mitchell A. Thornton, *Southern Methodist University*
Series ISSN
Synthesis Lectures on Digital Circuits and Systems
Print 1932-3166   Electronic 1932-3174

Synthesis Lectures on Digital Circuits and Systems

Editor
Mitchell A. Thornton, *Southern Methodist University*

The Synthesis Lectures on Digital Circuits and Systems series is comprised of 50- to 100-page books targeted for audience members with a wide-ranging background. The Lectures include topics that are of interest to students, professionals, and researchers in the area of design and analysis of digital circuits and systems. Each Lecture is self-contained and focuses on the background information required to understand the subject matter and practical case studies that illustrate applications. The format of a Lecture is structured such that each will be devoted to a specific topic in digital circuits and systems rather than a larger overview of several topics such as that found in a comprehensive handbook. The Lectures cover both well-established areas as well as newly developed or emerging material in digital circuits and systems design and analysis.

vi

Pragmatic Electrical Engineering: Systems and Instruments

William Eccles
Rose-Hulman Institute of Technology

SYNTHESIS LECTURES ON DIGITAL CIRCUITS AND SYSTEMS #34

ABSTRACT

Pragmatic Electrical Engineering: Systems and Instruments is about some of the non-energy parts of electrical systems, the parts that control things and measure physical parameters. The primary topics are control systems and their characterization, instrumentation, signals, and electromagnetic compatibility.

This text features a large number of completely worked examples to aid the reader in understanding how the various principles fit together.

While electric engineers may find this material useful as a review, engineers in other fields can use this short lecture text as a modest introduction to these non-energy parts of electrical systems. Some knowledge of basic d-c circuits and of phasors in the sinusoidal steady state is presumed.

KEYWORDS

electrical engineering, control systems, characterizing systems, instrumentation, bridge circuit, signal, filter, electromagnetic compatibility

Contents

Preface

Electrical engineering deals with "electricity" for both the movement of energy and the transmission of information. *Pragmatic Electrical Engineering: Systems and Instruments* introduces several topics from this non-energy side. I have had three goals in mind as I have written this:

- Give you some basic ability to handle electrical problems you might encounter in your own work

- Let you determine where your limits are and when to go for outside help

- Develop enough understanding of the topic so that, when a consultant proposes something, you are prepared to ask the right questions

Four topics from the non-energy side of electrical engineering are discussed in this text. The first two chapters are about control systems, taking you only far enough to understand what constitutes a control system, how it functions, and why it works. This includes a development of how to determine the basic parameters of a system that is to be controlled.

A chapter on instrumentation uses two sensors as the basis for examples, the strain gauge and the resistance temperature detector. These are introduced with the Wheatstone bridge as the basic circuit, along with the reasons why this bridge is used.

The chapter on analog signals focuses primarily on filtering signals. The goal is to learn how simple filters work, but more important, how their characteristics are represented.

Electromagnetic compatibility sounds like a very messy topic, and it really is! This chapter is about the basic principles of how electrical noise gets into a circuit and how one prevents one circuit from interfering with another.

If you are not reasonably familiar with electrical circuits, both basic direct-current circuits and also with phasors in sinusoidal steady-state circuits, you might want to study *Pragmatic Electrical Engineering: Fundamentals* as an introduction. The first two chapters of that text (which is another in this lecture series) should get you started.

These texts have all of the meanings of the word *Pragmatic* that appears in all their titles. They are my nonidealistic, practical, opinionated look at some topics in electrical engineering. Pragmatism is a way of learning where we understand first the practical side of things.

You won't turn into an electrical engineer through studying this text, but you might have a better understanding of systems and instruments in the field where you work. You probably have already discovered that you can't completely avoid "electricity!"

Wm. J. Eccles

May 2011

CHAPTER 1

Closed-Loop Control Systems

Control systems. That can mean all sorts of things, like the way a despot controls the vassals or the way a dispatcher controls trains. Let's narrow this down some. Control systems here mean systems that are designed to control some kind of process. The process may be just about anything; the control may be electrical or just about anything else. Hmmm…did that last sentence really say anything?

Let's start by thinking about open-loop systems. A very simple one is a doorbell. While you are controlling the doorbell, there is nothing you can do to make it louder or softer. If the bell is positioned so it can't be heard from the door, you can't even be sure it is ringing. In other words, there is no feedback, no guarantee of proper operation, or even operation itself.

Now how about closed-loop control systems? Basically, some information from the process being controlled comes back to the controller with the aim of improving what is happening. This "improving" might take the form of a more robust system because it continues to operate properly even in the face of perturbations. "Improving" might mean expanding the bandwidth of the process's response, making it a more responsive system to sudden changes.

You have participated in a closed-loop control system many times as you have steered your car. Your hands control the steering wheel, the steering wheel controls the front wheels, the front wheels determine the direction of motion, the motion is seen by your eyes, your eyes control your brain, and your brain controls your hands. Most of the time, this closed-loop system functions very well and the car goes exactly where you want it to go.

We need a couple of tools before we study control systems. While these tools are useful for other things as well, it's in this chapter that they are really needed. These tools are the concept of the *decibel*, dB for short, and a graphical presentation called the *Bode plot*.

1.1 dB

Writing dB is fun because you don't often get to write a "word" and put the capital letter at the end. The *B* is there to honor Alexander Graham Bell. If you don't know who he is, it's time to turn in your cell phone and go back to tin cans and a string!

1.1.1 BELS AND DECIBELS

We start with a "unit", which is unitless, called the *bel*, abbreviated *B*. This is a *power ratio* expressed as a base-10 logarithm:

$$bel = \log_{10} \frac{P_{out}}{P_{in}} \; \text{B}$$

Note that this is a *power ratio,* which means that the bel does not have any units—it's watts per watt, a ratio.

Somewhere early in the life of the bel, somebody decided that the bel didn't have a decent "range." A power ratio of 10, for example, is just 1 bel. Similarly, a power ratio of 2 is only 0.3 bel, while a ratio of 100 is just 2 bels.

That's probably why we use the *decibel,* which is the bel multiplied by 10:

$$decibel = 10 \log_{10} \frac{P_{out}}{P_{in}} \; \text{dB}$$

Sometimes we want dB to convey actual power information, not just a ratio. In these cases, we establish a reference. An example might be to set the reference to one milliwatt. Then an output power of 150 mW comes out as

$$P = 10 \log_{10} \frac{150}{1} = 21.8 \; \text{dBmW}$$

Some common power references relate to antennas. One is a comparison between the power that an antenna radiates in a certain direction compared with what it would radiate uniformly around a sphere. This is called dBi, or decibels isotropic. Another is dBd where the comparison is with a dipole antenna.

Now think back to what we have done with circuits. We usually ask questions about voltages and currents in the circuit, not power. If we do ask about power, we do this after we have found the voltages and currents.

The dB is also a useful quantity for comparing voltages. The catch is that the dB is defined for power, so we have to somehow relate voltage to power. We do this by establishing a reference impedance. This reference impedance can be any value. In some audio work, the reference is 600 Ω, for example. But as we will see shortly, it makes no difference what we choose as far as the calculations are concerned.

I want to use dB to compare voltages, so I would like to have a ratio of voltages. Consider the circuit of Fig. 1.1 that shows a reference resistor R_{ref} and an applied voltage V. Now suppose I reference two different voltages to that resistor. Each voltage delivers a certain power to the reference resistor:

$$P_1 = \frac{V_1^2}{R_{ref}}, \quad P_2 = \frac{V_2^2}{R_{ref}}$$

So their ratio in dB is

Figure 1.1: Reference

$$dB = 10\log_{10}\frac{V_2^2/R_{ref}}{V_1^2/R_{ref}}$$

But the reference resistance cancels out leaving the ratio of the squares of the voltages:

$$dB = 10\log_{10}\frac{V_2^2}{V_1^2}$$

The log of a square is two times the log:

$$dB = 20\log_{10}\frac{V_2}{V_1}$$

There! That's the dB we use. But it isn't the primary definition—the primary one is based on power.

There are some standard voltage references, just as there are for power. One fairly common one is a reference of one microvolt: $dB\mu V$. Another is for audio work: 0 dBu is the reference voltage produced by 1 mW delivered to 600 Ω.

1.1.2   EXAMPLE I—USING dB

Here is how decibels can be used for calculating power levels. Suppose we know the power delivered to a feedline by a certain transmitter, the fraction of that power lost in the feedline, the fraction of the power that makes it through the balun network at the antenna, and the effective amount of power radiated by the antenna.

All of these quantities are "gains" in the sense that the gain is the ratio of the output to the input power. If I have these numbers, then I multiply them to get the overall gain. In other words, overall gain = gain1 x gain2 x....

Decibels are logarithmic, so this multiplication of gains becomes addition in dB.

Suppose our transmitter-antenna system has the following characteristics:

- Transmitter output power: 2.5 kW

- Feedline loss: −1.8 dB

- Balun network loss: -0.4 dB

- Antenna effective gain: 21.8 dB

What is the effective radiated power (ERP) of this setup?

The gain from the transmitter to the antenna output is $-1.8 - 0.4 + 21.8 = 19.6$ dB. I'm dealing with power, so I use the "10" version of dB because that is for power:

$$19.6 \text{ dB} = 10\log_{10}\frac{P_{ant}}{P_{xmtr}} = 10\log_{10}\frac{P_{ant}}{2.5}$$

Now divide through by 10 and then raise 10 to the value on each side:

$$\frac{19.6}{10} = \log_{10}\frac{P_{ant}}{2.5}$$

$$10^{19.6/10} = 10^{\log_{10}\frac{P_{ant}}{2.5}}$$

But 10-to-the-log $_{10}$ leaves just the argument of the log:

$$10^{19.6/10} = \frac{P_{ant}}{2.5}$$

Solving for P_{ant} gives us the effective radiated power (ERP) of this transmitter-antenna system:

$$P_{ant} = 2.5 \times 10^{19.6/10} = 228 \text{ kW}$$

Don't worry! We aren't creating power by getting 228 kW ERP from a 2.5-kW transmitter. If you were to place your receiver right in the beam of the antenna, you'd think you were seeing 228 kW. But if you got, say, straight below the antenna, you'd think the antenna was radiating almost nothing. Averaged over a sphere around the antenna, the radiated power is about 1.5 kW.

1.1.3   EXAMPLE II—GAIN FROM dB

The system of Fig. 1.2 is described in terms of voltages and voltage gains, so I will use the "20" version of dB because that is for voltage. We are to find the input voltage that is needed to produce the output of 2.0 V.

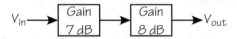

Figure 1.2: Use of dB

The overall gain of this system is $7 + 8 = 15$ dB, so the dB relationship here is

$$15 \text{ dB} = 20 \log_{10} \frac{2.0}{V_{in}}$$

Dividing by 20 and raising 10 to each side of the equation yields a result than can be solved for the voltage input:

$$10^{15/20} = 10^{\log_{10} 2.0/V_{in}}$$

$$V_{in} = \frac{2.0}{10^{15/20}} = 356 \text{ mV}$$

1.2   BODE PLOTS

Hendrik W. Bode (1905–1982) was a pioneer in both controls and telecommunications. Unfortunately, he has the kind of name that has several ways of being pronounced: Bôw'dee or Bôw'dah or Bôw'duh. Mr. Bode probably used the Dutch pronunciation, which rhymes with Yoda, although Yoda probably bears no resemblance to Mr. Bode.

Bode spent a good part of his career with the Bell Telephone Laboratories. He probably did a lot of work with signals and filters and communications lines. Doing this required him to plot the gain of a system, both magnitude and phase, against frequency. Plots of this nature are a very common way of visualizing how a filter handles the various frequencies of a signal.

So of course Bode used the most powerful computer he had available. He would write out the transfer function as a function of frequency and then have the computer plot two graphs, the magnitude of the transfer function and the phase angle of the transfer function, both against a frequency axis. He chose, like most practitioners of the day, to use a logarithmic frequency axis.

But there probably came a time when he got tired of doing all this computation. Even though he probably had one of the more advanced computers of the day, the work was still time-consuming. Remember that this was in the late 1930s.

And what was Bode's computer? Very likely a Keuffel&Esser Model 4080-3 Log-Log Duplex Trig sliderule, introduced by K&E in 1937. (You might have fun Googling it.) It was the top-of-the-line sliderule, although it was supplanted about ten years later with the Log-Log Duplex Decitrig rule that did angles in decimal degrees rather than degrees-minutes-seconds.

Bode developed the method of making the plot that carries his name to reduce the computation load on his sliderule. Today we use it mostly to display the magnitude and phase characteristics of a transfer function, get quick estimates of how a system might behave, and help determine how to adjust system response.

Bode's method is based on the ease by which the *asymptotes* of a magnitude or phase plot can be drawn if the vertical axis is in dB or degrees and the horizontal axis is the logarithm of frequency.

1.2.1   ASYMPTOTES MATH

Bode plots are for both magnitude and phase of transfer functions. We are going to restrict our look to just magnitudes. In other words, we are going to plot $|H(j\omega)|$ against $\log_{10}\omega$. We could be using hertzian frequency just as well, but we'll start with radians/second. For the magnitudes we will use dB.

While there are lots of possible configurations of $|H|$, we begin with just three:

$$H(s) = s$$

$$H(s) = 1 + \frac{s}{a}$$

$$H(s) = \frac{1}{1 + \dfrac{s}{a}}$$

Since $s = \sigma + j\omega$ and we are interested in just the sinusoidal steady-state response, we will restrict the frequency to $s = j\omega$ (or $s = j2\pi f$ for hertz):

$$|H(j\omega)| = \omega$$

$$|H(j\omega)| = \left|1 + \frac{j\omega}{a}\right|$$

$$|H(j\omega)| = \frac{1}{\left|1 + \dfrac{j\omega}{a}\right|}$$

Our asymptotic analysis will start by finding the magnitude of H in dB (voltage dB) and then seeing what the asymptotes look like on a graph of dB versus $\log\omega$.

From here on, I am going to omit the subscript "10" on the log symbol.

Convert the first of the H forms to dB:

$$20\log|H(j\omega)| = 20\log\omega$$

Now we reason our way through the curve of $|H|$ versus frequency:

- At $\omega = 1$, $|H| = 0$ dB

- At $\omega = 10$, $|H| = 20$ dB

- At $\omega = 100$, $|H| = 40$ dB

And so on. $20\log\omega$ defines a *straight line* whose slope is 20 dB for each step of 10 along the log-ω axis. This step of 10 is a decade. Figure 1.3 shows this asymptote.

In other words, if one of the factors of H(s) is just s, this factor is plotted as a straight line passing through ($\omega = 1$, dB$= 0$) and sloping upward at *20 dB per decade*. If one of the factors of H(s) is 1/s, the line passes through ($\omega = 1$, dB$= 0$) but slopes *downward* at 20 dB per decade.

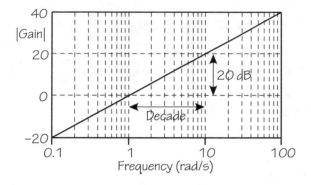

Figure 1.3: Bode plot for $|H(s)| = s$

Well, that wasn't too bad. How about the second factor where H(s) contains the factor 1 + s/a? Take a look at the following manipulation of this factor, first getting its magnitude, then its magnitude in dB as a function of ω.

$$H(j\omega) = 1 + \frac{j\omega}{a}$$

$$|H(j\omega)| = \sqrt{1 + \frac{\omega^2}{a^2}}$$

$$20\log|H(j\omega)| = 20\log\sqrt{1 + \frac{\omega^2}{a^2}} = \frac{20}{2}\log\left(1 + \frac{\omega^2}{a^2}\right)$$

$$= 10\log\left(1 + \frac{\omega^2}{a^2}\right) \text{ dB}$$

That looks messy, but let's not get eaten by the alligators while we are working in the swamp. Remember that our goal is to get the *asymptotes* of the function, not the exact path. So we'll look at what happens in two regions, one when ω is much smaller than a and getting smaller, the other when ω is much larger than a and getting larger.

As ω gets much smaller than a, our term becomes

$$\lim_{\omega \to 0}\left[|H(j\omega)|\right] = \lim_{\omega \to 0}\left[10\log\left(1 + \frac{\omega^2}{a^2}\right)\right]$$

$$= 10\log(1) = 0$$

This says that the asymptote for small values of ω is a horizontal line at 0 dB.

What does the asymptote in the other direction look like? Here's the math:

$$\lim_{\omega \to \infty}\left[\left|H(j\omega)\right|\right] = \lim_{\omega \to \infty}\left[10\log\left(1+\frac{\omega^2}{a^2}\right)\right]$$

$$= 10\log(\frac{\omega^2}{a^2}) = 20\log(\frac{\omega}{a})$$

Let's do the same thing we did with $20\log\omega$:

- At $\omega = a$, $|H| = 0$ dB

- At $\omega = 10a$, $|H| = 20$ dB

- At $\omega = 100a$, $|H| = 40$ dB

This defines a straight line whose upward slope is 20 dB per decade. Hmmm, haven't we just heard about one of those?

Putting this all together, if one of the factors of H(s) is $1 + s/a$, there are two asymptotes. One is a straight line from $\omega = a$ on the log-ω axis, heading to the left, at 0 dB. The other is a straight line starting at $\omega = a$ on the log-ω axis and heading upward at 20 dB per decade. The two asymptotes meet at $\omega = a$ on the log-ω axis. Figure 1.4 shows this asymptote.

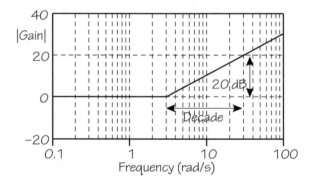

Figure 1.4: $|H(s)| = 1 + s/a$ for $a = 3$

Now for the third possible factor, where H(s) contains the factor 1 + s/a in its denominator, or in other words, the factor 1/(1 + s/a). I'm not going to work that out here because it is the same as the factor 1 + s/a, except that the asymptote starting at $\omega = a$ on the log-ω axis slopes *downward* at 20 dB per decade. Figure 1.5 shows this asymptote.

Now we have all the asymptotes for first-order terms for $|H(s)|$:

- $|H(s)| = s$: Line sloping upward at 20 dB per decade and passing through ($\omega = 1$,dB= 0). (Fig. 1.3)

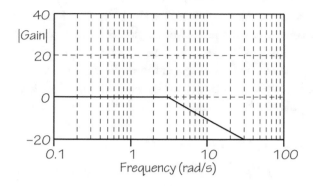

Figure 1.5: $|H(s)| = 1/(1 + s/a)$ for $a = 3$

- $|H(s)| = 1 + s/a$: Two lines, one starting at $\omega = a$ and going to the left at 0 dB; the other starting at $\omega = a$ and sloping upward to the right at 20 dB per decade. (Fig. 1.4)

- $|H(s)| = 1/(1 + s/a)$: Two lines, one starting at $\omega = a$ and going to the left at 0 dB; the other starting at $\omega = a$ and sloping downward to the right at 20 dB per decade. (Fig. 1.5)

One final remark on these slopes: *Only* 20 dB per decade and integral multiples are possible. That means ± 20 or ± 40 or ± 60 or ± 80 and so on, but not 15 or 23.98 or anything else.

It is sometimes convenient to state these slopes in *dB per octave*. 20 dB per decade is the same as 6 dB per octave, where an *octave* is a factor of 2.

1.2.2 PUTTING THE ASYMPTOTES TOGETHER

Gosh! I now know all about asymptotes and I can draw all sorts of lines that show me something about factors of $|H(s)|$. That sounds like something I've always wanted to…. Um, I'm not sure that's true. So, OK, why do I need those asymptotes?

Well, consider. Suppose I have a system, described by H(s) in one of its various forms like H(s) or H(jω) or H(j2πf). It'll look something line this:

$$H(s) = \frac{(A\ constant)(Some\ factor\ in\ s)}{(Another\ s\ term)(A\ third\ one)}$$

This H(s) is a combination of factors. The first is divided by the other two. Those bottom two are multiplied together. Now remember what logs do for us. Multiplication becomes addition; division becomes subtraction. So this H(s), when converted to $|H(j\omega)|$ and then converted to dB (logs, that is) becomes

$$\log|H(j\omega)| = \log(A\ constant) + \log|Some\ factor\ in\ s|$$
$$- \log|Another\ s\ term| - \log|A\ third\ one|$$

That says that we create the asymptotes for the individual factors, one by one, and then we add or subtract as appropriate to get the final Bode plot. So if you were worried how this gets back to Bode and the plots, that is how it'll appear. (OK, so you weren't worried….)

Actually, it's even easier. If we have the asymptotes for the various factors, we just add them on the graph—no subtraction. Why no subtraction? Because we have already developed the asymptotes for a factor in the denominator that includes the sign—that's where we got the downward-at-20-db/decade slope.

The basic algorithm for generating Bode plots:

- Rearrange H(s) to put all the factors in the form of either s or $1 + s/a$

- Draw the asymptote for the constant term, if any, converted to dB (a horizontal straight line)

- Draw the asymptote for the s term, if any, passing through $\omega = 1$ at 0 dB. The line slopes upward at 20 dB per decade if the s is in the numerator, downward if it's in the denominator

- Draw the asymptotes for any factors in the numerator by drawing a horizontal line from $\omega = a$ to the left at 0 dB and a second line with an upward slope of 20 dB per decade from $\omega = a$.

- Do the same thing for any factors in the denominator, but sloping downward at 20 dB per decade

- Add these asymptotes graphically

Now some examples.

1.2.3   EXAMPLE III—BODE PLOT FOR ONE POLE

Plot the Bode magnitude plot for

$$H(s) = \frac{4000}{s + 2000}$$

First, convert this to the 1 + s/a form by dividing through by 2,000:

$$H(s) = \frac{2}{1 + \dfrac{s}{2000}}$$

Now convert the numerator factor (2) to dB:

$$20 \log 2 = 6 \text{ dB}$$

Draw this on the plot (Fig. 1.6, dotted line at 6 dB).

Now plot the asymptotes of the denominator, which are a line to the left from $\omega = 2,000$ at 0 dB and a line sloping down to the right, starting at $\omega = 2,000$, with a slope of -20 dB per decade (Fig. 1.6, dotted line).

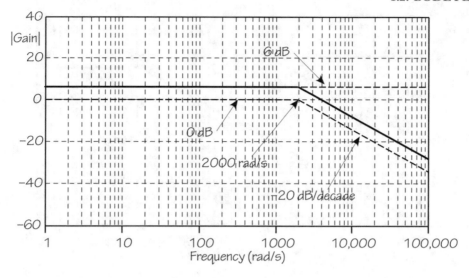

Figure 1.6: Example III

Sum the asymptotes. To the left of $\omega = 2,000$, $0 + 6 = 6$ dB. To the right, move the downward slope up by 6 dB. The result is the solid lines of Fig. 1.6.

1.2.4 EXAMPLE IV—BODE PLOT FOR POLE AND ZERO

Plot the asymptotes for the magnitude of this H(s):

$$H(s) = \frac{5s}{s + 500}$$

Convert:

$$H(s) = \frac{\dfrac{5}{500}s}{1 + \dfrac{s}{500}}$$

The d-c constant is $5/500 = 0.01$, which in dB is -40 dB. So this asymptote is a horizontal line at -40 dB (Fig. 1.7).

The s in the numerator is a line sloping upward to the right and passing through $\omega = 1$ and 0 dB (Fig. 1.7).

The asymptotes for the denominator factor are two lines. One starts at $\omega = 500$ and goes horizontally to the left at 0 dB. The other starts at 500 and goes downward to the right at -20 dB per decade (Fig. 1.7).

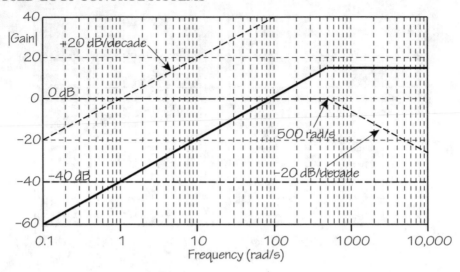

Figure 1.7: Example IV

Now add them, working from left to right to get the result shown in Fig. 1.7. Notice that adding a slope of +20 dB per decade to a slope of −20 dB per decade yields a flat line (a slope of 0 dB per decade).

1.2.5 EXAMPLE V—BODE PLOT FOR TWO POLES

Given

$$H(s) = \frac{7.5 \times 10^4 s}{(s+150)(s+2500)}$$

Convert to

$$H(s) = \frac{0.2s}{\left(1+\dfrac{s}{150}\right)\left(1+\dfrac{s}{2500}\right)}$$

The numerator constant is 0.2, which is −14 dB, which is a horizontal line at −14 dB. (Fig. 1.8)

The numerator factor s is an upward slope of 20 dB per decade passing through $\omega = 1$ and 0 dB.

The denominator has two factors. One breaks at $\omega = 150$ and the other at $\omega = 2,500$. Their flat parts are along 0 dB. One breaks downward at −20 dB per decade at $\omega = 150$ and the other breaks downward at $\omega = 2,500$.

The sum of these finishes the plot in Fig. 1.8.

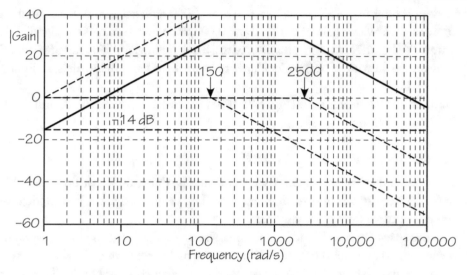

Figure 1.8: Example V

1.2.6   EXAMPLE VI—BODE PLOT TO TRY

This H(s) has an *s* in the denominator:

$$H(s) = \frac{18\left(1 + \dfrac{s}{200}\right)}{s\left(1 + \dfrac{s}{3000}\right)}$$

It's already converted to proper form.

The constant term is 18, which is $20 \log 18 = 25$ dB.

The numerator's factor is level at 0 dB to the left of $\omega = 200$ and slopes upward at 20 dB per decade to the right of $\omega = 200$.

The *s* term in the denominator is a downward slope of -20 dB per decade passing through $\omega = 1$ and 0 dB.

The denominator term breaks downward to the right at $\omega = 3,000$.

Their sum is…left as an exercise for you. It passes through ($\omega = 0$, 25 dB) sloping downward to $\omega = 200$, flattens out at -21 dB, then continues the downward slope starting at $\omega = 3,000$.

1.2.7   WHAT DO THESE MEAN?

The four examples we've just done describe three different types of filters. While we haven't proven it, the *break frequency* of a circuit is the "joint" between the asymptotes. So in Example III (Fig. 1.6), we have a low-pass filter with a break frequency of $\omega = 2,000$ radians/second. Below that frequency, the filter passes signals with a gain of 6 dB. Above 2,000, it begins to cut off the signal.

Example IV (Fig. 1.7) is a high-pass filter with a break frequency of 500 radians/second. Above that, it passes the signal with what the plot shows to be a 14-dB gain. Below that, it diminishes the signal.

Example V is a band-pass filter (Fig. 1.8). It has a pass-band gain of 27 dB and breaks at 150 and 2,500 radians per second.

1.2.8   FINISHING UP BODE'S WORK

Hendrik Bode didn't do this just for the asymptotes, and he didn't do it just for first-order factors. I'm leaving the higher-order factors to another course, but we do need to see how Bode constructed the actual response curve from the asymptotes.

Part of the answer of how he did it is obvious: the curve must conform to the asymptotes. But he devised a scheme that enabled him to sketch in the actual curve within a precision of about half a dB of the exact curve that he could have calculated using his Log-Log Duplex Trig sliderule.

Here are the steps, which I'll draw on the Bode plot of Example V. (See Fig. 1.9.)

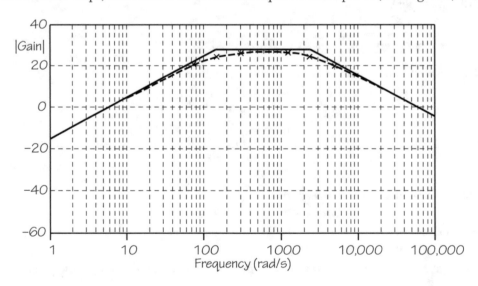

Figure 1.9: Example V complete

- *Inside* the corners at $\omega = 150$ and $\omega = 2,500$, place a point 3 dB inside the corner. That means dropping down 3 dB from the corner to mark the point.

- Now let's work near the corner at $\omega = 150$. From this corner, move to the left by a factor of two, which puts us at $\omega = 75$. Place a point 1 dB *inside* the curve.

- Likewise, move from this corner to the right by a factor of two (i.e., $\omega = 300$ and place a point 1 dB inside the curve.

- Do the same for the other corner at $\omega = 2,500$, moving down to 1,250 and up to 5,000.

- Now eyeball a smooth curve following the asymptotes but passing through the six points that you marked.

If you have drawn carefully and then compare with results from Maple or Matlab, you should have trouble telling them apart. (Figure 1.9)

That's Bode's genius.

1.3   CLOSED-LOOP SYSTEMS

One way to start looking at closed-loop control systems is to look at open-loop systems first. Doing this might give us some perspective on the why of closed-loop.

The system of Fig. 1.10 is an open-loop control system. The input to the system is r(t). That

Figure 1.10: Open-loop, t domain

input is to produce a desired outcome, y(t). An example might be a system (the G block) consisting of a power amplifier and a d-c motor. A voltage r(t) drives the input of the amplifier. The amplifier amplifies this voltage to make the d-c motor run. The output y(t) is the desired rpm of the motor.

This turns out to be a not-so-hot control system! It does control the speed of the motor but not very well. For example, suppose there is a linear relation between the input voltage and the rpm of the motor. To keep things simple, one volt at the input r(t) makes the motor output $y(t) = 1,000$ rpm.

Now add a load to the motor. It naturally slows down, so y(t) is less than 1,000 rpm. But r(t) is still 1 volt, so we think the motor should still be producing 1,000 rpm. Our system is not *robust* in that it cannot respond to changes at the output.

So far, I'm thinking in the time domain with my variables as functions of t. We won't continue in the time domain but instead will switch to the *frequency* or *phasor* domain and consider operation just in the sinusoidal steady state. Figure 1.11 shows our simple open-loop system in phasor domain.

Now R(s) is the reference signal and Y(s) is the output. The amplifier and the motor in our example are G(s). But switching to the phasor domain doesn't do a thing to improve our system. It is still open-loop and the input reference is not aware of any changes in the output. It is still not robust.

It's time to close the loop.

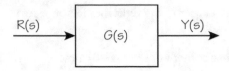

Figure 1.11: Open-loop, *s* domain

1.3.1   CLOSING THE LOOP

Closed-loop systems close the loop by watching the output, then feeding back that information to the input. What comes back is compared with the reference. If they don't match, the system is adjusted to make them match. Now our system can be robust, because a change in the output changes the input to compensate for the change in the output.

Figure 1.12 shows the basic closed-loop system. Here's what the various pieces do:

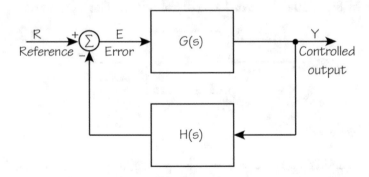

Figure 1.12: Closed-loop system

$G(s)$ is the *forward* path. It is made up of the plant (like the d-c motor of my open-loop example) and the driver for that plant (like the amplifier in my example).

$H(s)$ is the *feedback* device that gets information from the output and delivers it to be compared with the input (like a tachometer).

$R(s)$ is the *reference* signal that establishes what we are trying to do (like the input voltage in my example).

$Y(s)$ is the *output* of the system (like the rpm of the d-c motor).

$E(s)$ is the *error* signal, which is the difference between the reference R(s) and the actual output Y(s).

The summer is the comparator, subtracting the feedback information from the reference to create the error signal E(s).

The two blocks in the system, G(s) and H(s) are not functions of time—they stay constant as far as we are concerned. Yet this is not 100% true. They perhaps drift slowly with time or temperature or voltage change. For our analysis, though, we consider them to be *static,* meaning not time-varying.

If anything in this system needs to be stable and predictable, it is H(s). It is important that H(s) observes the output and reports the results to the summer in a very consistent manner. We will take some care to be sure H(s) in the feedback path can be expected to behave.

1.3.2 ANALYZING THE LOOP

The output Y(s) depends on the error signal E(s) driving G(s):

$$Y(s) = G(s)E(s)$$

The error signal E(s) is the difference between the reference R(s) and the signal fed back from the output:

$$E(s) = R(s) - H(s)Y(s)$$

Combining these two equations yields the classic description of a feedback system:

$$Y(s) = \frac{G(s)}{1 + G(s)H(s)} R(s)$$

Everyone who works with closed-loop systems knows this one by heart! One way of stating it mentally is, "Forward gain over 1 plus loop gain." The forward gain is the path from left to right, namely, G(s). The loop gain is clockwise around the loop, namely, G(s)H(s). Another common way to say this is, "G over 1 plus GH."

Now let's examine the characteristics of the two big players in this equation, G(s) and H(s). We can make the following generalizations, that are *usually* true in a control system:

- G(s) is usually a heck of a lot bigger than 1. How "heck of a lot"? Back to my motor example. A small error voltage must drive a large motor, so the gain needed to do this is large. We can write something like G >>> 1.

- H(s) is generally not larger than 1. While it can be 1, meaning that the output is fed back without change, bigger than 1 is bad. Why? Because this implies an amplifier, which has inaccuracies like long-term drift and change due to voltage supply changes and so on. We want H(s) to be as perfect as possible because it is measuring how well our reference is controlling the output. So H < 1.

Now I'm a bit stuck for notation. I've used R and E and G and Y and H, which is pretty common but by no means universal. However, H is now "used up" and I can't use it for labeling the overall transfer function as in previous chapters. I'm going to use TF for "transfer function" to mean the overall relationship between output and input:

$$TF(s) = \frac{G(s)}{1 + G(s)H(s)}$$

If G >>> 1 and H < 1, look what happens to TF:

$$TF = \frac{G}{1 + GH}$$

for $G >>> 1$ and $H < 1$, $GH >> 1$

$$TF \cong \frac{G}{GH} = \frac{1}{H}$$

In other words, the ideal, perfect system has a transfer function that is simply 1/H. But if the system is perfect, it can't work! We can see this by figuring out what the error signal E(s) will be under these perfect conditions:

$$E = R - H\frac{R}{H} = 0 \text{ for } Y = \frac{1}{H}R$$

The error signal E(s) drives the amplifier or whatever is in the forward path. If E(s) is zero, it can't drive anything. If nothing is driven, there is no output. So the system can't work. How close can we get to "perfect?" We need to look at error in the desired output Y(s).

1.3.3   EXAMPLE VII—TRANSFER FUNCTION

A certain plant requires a gain of 40 dB $\pm$ 5%, which is a gain of between 38 and 42 dB. This means that the reference signal R(s) is to be amplified by 40 dB to produce the output Y(s). The amplifier in the forward path is typical of amplifiers: its gain is not "perfect" but varies from 10^4 to 10^5. (Forward gain is often called the *open-loop gain*.)

First we consider the ideal situation where $TF(s) = 1/H(s)$:

$$TF = 40 \text{ dB } = 10^{40/20} = 100$$

$$H(s) = \frac{1}{TF(s)} = 0.01$$

Now find the system gain TF(s) at the ends of the range of amplifier gain:

for $G(s) = 10^4$

$$TF(s) = \frac{10^4}{1 + (10^4)(0.01)} = 99$$

for $G(s) = 10^5$

$$TF(s) = \frac{10^5}{1 + (10^5)(0.01)} = 99.9$$

The system gain is supposed to be 100. Depending on the actual gain of the amplifier, the system gain is really between 99 and 99.9. That's an error of between 1% and 0.1%. In dB, this range is 39.91 to 39.99 dB, well within the 40±5% specification. The system won't be perfect until G is infinite!

1.3.4   FRACTIONAL ERROR

The ideal or perfect system has a transfer function of 1/H(s). But we can't have that and instead have some less-than-perfect transfer function TF(s). We define the *fractional error* as

$$E_{frac} = \frac{\dfrac{1}{H(s)} - TF(s)}{TF(s)}$$

$$= \frac{\dfrac{1}{H(s)} - \dfrac{G(s)}{1 + G(s)H(s)}}{\dfrac{G(s)}{1 + G(s)H(s)}} = \frac{1}{G(s)H(s)}$$

Fractional error is just 1 over the loop gain.

1.3.5   EXAMPLE VII CONTINUED

For $G = 10^4$, the gain was 99.0, an error of 1%. The fractional error is

$$E_{frac} = \frac{1}{(10^4)(0.01)} = 0.01$$

which is 1%. For $G = 10^5$, the gain was 99.9, an error of 0.1%. The fractional error is

$$E_{frac} = \frac{1}{(10^5)(0.01)} = 0.001$$

which is 0.1%.

Suppose G falls to 10^3. The fractional error is

$$E_{frac} = \frac{1}{(10^3)(0.01)} = 0.1$$

which is 10%.

1.3.6   EXAMPLE VIII—MOTOR SPEED CONTROL

The d-c motor of Fig. 1.13 is to produce 2,000 rpm with a control voltage of 2 volts. The motor's

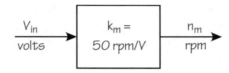

Figure 1.13: D-c motor

response is linear, so a control voltage of 0 to 2 volts is to produce 0 to 2,000 rpm. The motor constant is $k_m = 50$ rpm/V.

If we do this job using an open-loop system, we need an amplifier with a gain of A (see Fig. 1.14):

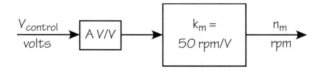

Figure 1.14: D-c motor: open-loop control

$$V_C = 2 \text{ V} \implies n_m = 2000 \text{ rpm}$$
$$n_m = (A)(50)V_C$$
$$A = 20$$

The open-loop system needs a gain of 20 to drive the motor so 2 volts input produces 2,000 rpm. But what happens to the motor speed when the load on the motor increases? Very simply, the speed just goes down. The control voltage is still 2 V, but the speed is no longer 2,000 rpm.

Well, how about the closed-loop control system in Fig. 1.15? (Note the use of units in the blocks—it's not a bad idea to do this.)

The tachometer must "match" the output to the input. V_C at 2 volts corresponds to $n_m = 2,000$ rpm. This means the tachometer constant must be $k_t = 2/2000 = 0.001$ V/rpm. This meets our requirement that H(s) be no larger than 1.

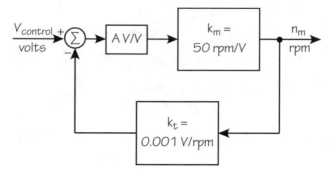

Figure 1.15: D-c motor: closed-loop control

That takes care of H(s). Now how about G(s)? It's tempting to say that we already have, because in the open-loop system in Fig 1.3-5, the amplifier and the motor together have a gain $G(s) = (20)(50) = 1,000$ rpm/V. That seems perfect because 2 V in gives 2,000 rpm out.

But the amplifier is no longer amplifying the input voltage V_c; it's amplifying the error voltage from the summer. If we want "perfect" output, meaning exactly 2,000 rpm for 2 V in, we need the amplifier to have infinite gain. How do I get this conclusion? Look at the fractional error:

$$E_{frac} = \frac{1}{GH} = \frac{1}{0.001G(s)}$$

For the fractional error to be 0 (i.e., perfection), G(s) must be infinite. So the next question is, how much error are you willing to accept in your system?

Let's say we'll accept 2% fractional error:

$$E_{frac} = 0.02 = \frac{1}{G(s)H(s)}$$

$$G(s) = \frac{1}{(0.02)(0.001)} = 5 \times 10^4$$

$$= Ak_m = 50A$$

$$A = 1000 \text{ V/V}$$

The actual maximum rpm for $V_c = 2$ V will be

$$n_m = 2\frac{(1000)(50)}{1+(1000)(50)(0.001)} = 1961 \text{ rpm}$$

Instead of 2,000 rpm, we get 1,961 rpm. Hmmm, maybe we want better than that. How about 1% fractional error?

$$E_{frac} = 0.01 = \frac{1}{G(s)H(s)}$$

$$G(s) = \frac{1}{(0.01)(0.001)} = 10^5$$

$$= Ak_m = 50A$$

$$A = 2000 \text{ V/V}$$

The outcome depends on what error you'll accept and how much gain you are willing to pay for in the amplifier. Notice that as the gain goes up, the fractional error improves.

1.3.7   EXAMPLE IX—LINEAR POSITIONER

The linear positioner shown in Fig. 1.16 can position an object over a range of 400 mm under control

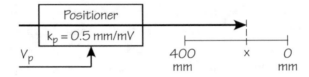

Figure 1.16: Positioner to be controlled

of a voltage ranging from 0 to 4 V. The positioner's constant is

$$k_p = 0.5 \text{ mm/mV}$$

The position x is to be precise to within ± 0.5 mm.

Two laser range finders are available with sufficient precision. Each produces a voltage proportional to distance:

$$\#1: \ V_f = 2 \text{ mV/mm}$$

$$\#2: \ V_f = 20 \text{ mV/mm}$$

Range finder #1 is less expensive, so let's try to design with it first. The maximum output voltage of range finder #1 will be

$$V_f = \left(2 \ \frac{mV}{mm}\right)(400 \text{ mm}) = 800 \text{ mV}$$

Hmmm, that should be OK. We can just put a gain of 5 in the feedback path to get the proper comparison with the 4-V control voltage.

No way! We must not do this! Putting gain in the feedback path means sacrificing the precision of that path. Remember that the feedback should be as precise as possible and should be no greater than 1.

Let's try again with range finder #2:

$$V_f = \left(20 \ \frac{\text{mV}}{\text{mm}}\right)(400 \ \text{mm}) = 8 \ \text{V}$$

That's twice as big as we need. But it's OK to reduce that output because it can be done with passive components. A simple voltage divider made of precision resistors will probably do the job.
The complete system is shown in Fig. 1.17. Our final job is to select the gain A of the amplifier.

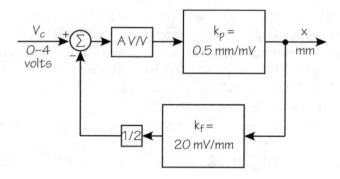

Figure 1.17: Positioner controller

We do this by making the fractional error fit the original specifications, which said we wanted a precision of no worse than ±0.5 mm.

$$E_{frac} = \frac{0.5 \ \text{mm}}{400 \ \text{mm}} = \frac{1}{G(s)H(s)}$$

$$= \frac{1}{(A)(0.5)(20)\left(\frac{1}{2}\right)}$$

$$A = 160$$

I'll check this result at V_c and see if it produces a value of x within ±0.5 mm of 400 mm:

$$G(s) = (160)(0.5) = 80 \ \text{mm/mV}$$

$$H(s) = (20)\left(\frac{1}{2}\right) = 10 \ \text{mV/mm}$$

$$TF(s) = \frac{80}{1+(80)(10)} = 0.099875 \ \text{mm/mV}$$

At the full control voltage $V_c = 4.000$ V:

$$x = (0.099875)(4000) = 399.5 \ \text{mm}$$

Great! The error is exactly 0.5 mm. Also note that

$$G(s) = 80 \quad \frac{1}{H(s)} = 0.1$$

I'll be adding to this example in the next section.

1.4   BANDWIDTH

The example of the positioner that I've just done isn't very realistic. Aw, gee, and you thought all the examples in textbooks were real? Sorry to disappoint you…. What's wrong with this one?

Think about how a positioner might work. Can it move instantly from one position to another? Does it perhaps dilly-dally? How does it really respond in the world of mass and dampers and finite forces and the like? Right! It can't act instantly. So our model of the positioner as just a box with a constant k_p doesn't describe the device very well.

Suppose this positioner is a first-order system. (It's really probably second-order, but let's keep this fairly simple.) As a linear first-order system, its response to a step is an exponential. Figure 1.18 shows a possible response. The time-domain equation for this response to a step $v_{in}(t)$ is

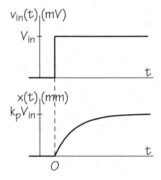

Figure 1.18: Physical system response

$$x(t) = k_p v_{in}(t)\left(1 - e^{-t/\tau}\right)$$

Take the Laplace transform of this to get

$$X(s) = \frac{k_p V_{in}}{s} - \frac{\tau k_p V_{in}}{\tau s + 1} = k_p \frac{V_{in}}{s}\frac{1}{\tau s + 1}$$

$$\frac{X(s)}{V_{in}/s} = \frac{k_p}{\tau s + 1}$$

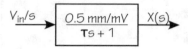

Figure 1.19: Realistic system

The positioner block in our system diagram now includes this first-order characteristic. Figure 1.19 is more realistic. Note that it still will have a d-c response (i.e., $s = 0$) of 0.5 mm/mV. That hasn't changed. Just the time it takes to actually move to a position has been added.

1.4.1 EXAMPLE IX CONTINUED

To get the time response of a device, we must characterize it from physical measurements, something we'll do in Chapter 2. Figure 1.20 is an example of a graph, taken from an oscilloscope, that shows how my positioner responds to a step input.

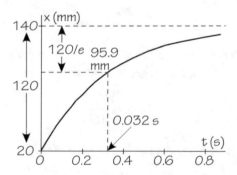

Figure 1.20: Characterizing

Remember that the time constant is the point on the exponential where the curve has 1/e to go to reach its final value. In this graph, the total excursion is 120 mm, so $1/e$ yields

$$\frac{1}{e}120 = 44.1 \text{ mm to go}$$
$$\text{so } x = 140 - 44.1 = 95.9 \text{ mm}$$
$$\text{when } t = \tau = 32 \text{ ms}$$

The new block diagram for the positioner is Fig. 1.21. We need the new transfer function TF(s) for this system. But before doing this for the positioner, we need a brief excursion into a way to simplify our algebra.

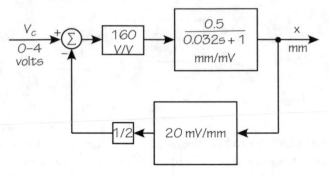

Figure 1.21: Positioner controller #2

1.4.2   SIMPLIFYING TF(s)

We already know that the transfer function of a closed-loop system is

$$TF(s) = \frac{G(s)}{1 + G(s)H}$$

First, note that H is generally not a function of s—it's just a number. Now, if we split G(s) into a numerator and a denominator, things get less messy:

$$G(s) = \frac{N(s)}{D(s)}$$

The result is a simpler way to do the algebra to get TF(s):

$$TF(s) = \frac{\dfrac{N(s)}{D(s)}}{1 + \dfrac{N(s)}{D(s)} H}$$

$$= \frac{N(s)}{D(s) + H \cdot N(s)}$$

Now back to the example.

1.4.3   EXAMPLE IX CONTINUED SOME MORE

For our improved system,

$$G(s) = \frac{80 \text{ mm/mV}}{0.032s + 1}, \quad H = 10 \text{ mV/mm}$$

Using the simpler way of getting TF(s) yields

$$N(s) = 80, \; D(s) = 0.032s + 1$$

$$TF(s) = \frac{80}{0.032s + 1 + (10)(80)}$$

$$= \frac{80}{0.032s + 801}$$

We need to present functions like these in standard form, which means that the polynomials in s have "1" as the coefficient of the highest-order s term. Doing that here gives us the standard form for TF(s):

$$TF(s) = \frac{0.099875}{1 + 39.95 \times 10^{-6} s} \; \frac{mm}{mV}$$

Check this result for d-c to see if it gives the same result as before. At d-c, $s = 0$, so the transfer function is

$$\text{at d-c, } TF(0) = 0.099875 \; \frac{mm}{mV}$$

OK! It came out the same. That helps to check our work.

1.4.4   FASTER RESPONSE

Something else has happened, though, that is very important to notice. The original positioner, without the closed-loop control system, has a time constant of 32 ms. That's what we measured when we characterized the positioner.

Take a look at the result of TF(s) after we have built the positioner into the closed-loop control system. The time constant of the actual response is

$$\tau = 39.95 \times 10^{-6} \cong 40 \; \mu s$$

FEEDBACK SPEEDS UP THE RESPONSE!

1.4.5   BODE REPRESENTATION

It turns out to be useful to plot on a Bode diagram the functions G(s), 1/H, and TF(s). I'll plot them as functions of $s = j\omega$ in Fig. 1.22.

First let's do G(s),

$$G(s) = \frac{80}{1 + 0.032s}$$

- At d-c, $G(0) = 80$, which is $20 \log_{10} 80 = 38$ dB, so the plot starts flat at 38 dB.

- The break frequency is at $\omega_b = 1/0.032 = 31.25$ radians/second, so the plot of G(s) breaks downward at 20 dB/decade at ω_b

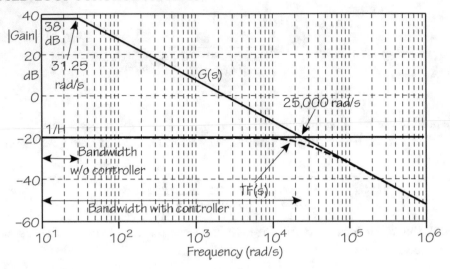

Figure 1.22: Bode diagram for positioner controller #2

Now let's do 1/H. Wait! you say. Why 1/H and not just H? Hold on for a bit and you'll see, but here's a clue: recall that the "perfect" system response is 1/H if G is very very large. Since $H = 10$ mV/mm, $1/H = 0.1$, which is -20 dB. (Note the minus sign.) This is a straight line on the Bode diagram.

From the Bode diagram we now have, we can deduce a few things:

- For $\omega < 10{,}000$ radians/second, $|G| \gg 1/H$. We can see that on the plot, because the G asymptote is 58 dB above 1/H. Under this condition, the actual transfer function TF(s) is very closely represented by just 1/H.

- For $\omega > 100{,}000$ radians/second, $|G| \ll 1/H$. Again, we can see the dB difference on the plot. Under this condition, the actual transfer function TF(s) is closely represented by G(s) alone.

- The dotted line on the Bode plot of Fig. 1.21 is the result of combining the trace of 1/H below where they cross and the trace of G(s) above where they cross.

- The break frequency of the complete closed-loop control system is where the two asymptotes cross at 25,000 radians/second (where the actual curve is 3 dB inside the corner).

Bandwidth is the frequency from 0 to the break frequency. Look what has happened to the bandwidth of the system:

- The original G(s) breaks at 31.25 radians/second, so its open-loop bandwidth is 31.25 radians per second.

• The closed-loop system breaks at 25,000 radians/second, so its bandwidth is increased to 25,000 radians per second.

FEEDBACK INCREASES SYSTEM BANDWIDTH!

The new bandwidth is defined by where G crosses 1/H: $\omega_{new} = \omega_{old}(1 + G_{dc}H)$.

1.5   STABILITY

We are not going to worry much about system stability in this course except to consider that systems can be designed and built that are not stable. What does *stable* mean? Firm, steadfast. Hmm, we need more than that.

A system is stable if its poles are in the left half plane. Ugh. OK, what's that saying? The poles are the roots of the denominator of the transfer function TF(s). Those roots must be to the left of the imaginary axis on the complex plane.

The denominator of the example we've just been working through is

$$1 + 0.032s$$

Setting that to zero gives is the position of this pole:

$$1 + 0.032s = 0$$

$$s = -31.25 \text{ s}^{-1}$$

The result is a negative value, so this root is in the left half of the complex plane and lies on the real axis at -31.25 seconds $^{-1}$

If the roots happened to be complex values such as for a quadratic, the real part of each of the roots must be negative.

There is much more to stability, including questions of stable but dangerous. For example, if you car has bad front shocks and you hit a chuck hole with one front tire, the front end steering and suspension may go into violent oscillation. The mechanical system is technically stable if the oscillation damps out, but the vehicle is unsafe nevertheless.

1.6   MORE EXAMPLE

We really haven't finished Example IX yet, but I'll cheat and finish it as Example X. Also, we haven't looked at how one might select a reasonable gain when designing a control system. Although we now know that "perfect" response comes with a gain G of infinity, we know we can't do that!

1.6.1   EXAMPLE X—REALLY IX CONTINUED AGAIN

We started Example IX with a positioner that moved instantly from one position to another. Then we gave the positioner a first-order exponential response. We saw that the feedback control made the system respond much faster and its bandwidth was much higher. "Faster" and "higher" go hand-in-hand.

We neglected the amplifier, though, by assuming that it just amplifies. Which of course it does. But not instantly. It like the positioner cannot go instantly from one output to another. So let's assume that, in its simplest form, it too follows an exponential trajectory.

I'm going to change the amplifier to one with a cutoff frequency of 100 radians/second, which could represent a heavily designed, elderly power amplifier. (That's 16 Hz, which is very low!)

If the cutoff frequency is 100 radians/second, the time constant of this first-order system is

$$\tau = 1/100 = 10 \text{ ms}$$

Compare 10 ms with the 32-ms time constant of the positioner and note that the amplifier is "faster" in responding than then positioner. This might imply that the amplifier won't be much of a factor in how our system responds.

Figure 1.23 is the new amplifier in the system. Calculations to get the transfer function produce

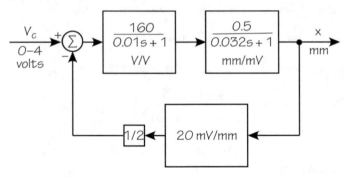

Figure 1.23: Positioner controller #3

a quadratic in the denominator this time:

$$G(s) = \frac{160}{0.01s+1} \frac{0.5}{0.032s+1}$$

$$= \frac{80}{0.00032s^2 + 0.042s + 1}$$

$$TF(s) = \frac{N_G}{D_G + H \cdot N_G}$$

$$= \frac{0.099875}{0.3995 \times 10^{-6} s^2 + 52.43 \times 10^{-6} s + 1}$$

Let's see how this second-order system behaves. We'll do this in more detail in Chapter 2. Doing this will show that the damping reatio is 0.0415, way less than 0.5. That means the system is not only underdamped, it is close to oscillatory.

That will yield very poor operation because the positioner's arm will overshoot its goal and then oscillate numerous times before settling down to where it belongs. There's just not enough damping.

The Bode diagram for this system (Fig. 1.24) is instructive. The d-c value is still 80. The plot

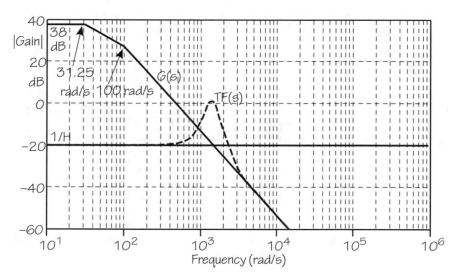

Figure 1.24: Too much underdamping

of 1/H has not changed. There are now two break frequencies:

$$\omega_1 = \frac{1}{0.032} = 31.25 \text{ rad/s}$$

$$\omega_2 = \frac{1}{100} = 100 \text{ rad/s}$$

(In order to get the peak of the curve as shown, I had to calculate the actual value of |TF| at several points near the crossing.)

I guess we had better try a faster amplifier (Fig. 1.25):

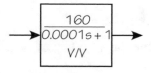

Figure 1.25: Another amplifier

$$\tau = 100 \ \mu s$$

$$\omega_b = 10,000 \ \text{rad/s}$$

The new Bode diagram is Fig. 1.26. Following the same analysis as before, we get a value for

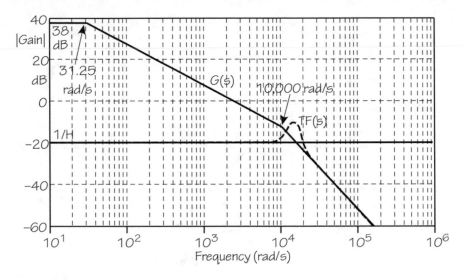

Figure 1.26: Still too much underdamping

the damping ratio (0.32) that is still too small for "decent" behavior.

OK, I'll try once more with an amplifier with a cutoff frequency of 25,000 radians/second (about 4 kHz). The time constant is 40 μs. Figure 1.27 is the block for this newest amplifier. The Bode diagram of Fig. 1.28) shows that the peak where |G| crosses 1/H is much smaller and the damping ratio is 0.5.

Actually, this system is right on the edge of being acceptable. "Acceptability" for second-order (and higher) systems is defined here as having a damping factor between 0.5 and 1. We can use the Bode diagram to get meet this requirement:

Here is a general rule for "acceptable," said two different ways:

• The 1/H line crosses the line for |G| before the second break frequency.

• 1/H crosses |G| before the second dive.

In this example, the crossing is right at the break that starts the last dive.

1.6.2 EXAMPLE XI—FINDING MAXIMUM GAIN

This is an example of how we can choose the forward gain from a Bode construction. Suppose my system is third-order:

Figure 1.27: Proper amplifier

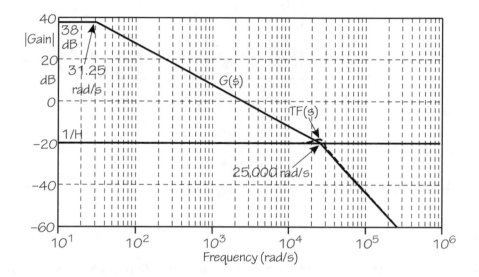

Figure 1.28: Borderline underdamping

$$G(s) = \frac{200k}{(1+0.01s)(1+0.05s)(1+0.2s)}$$

$$H = 0.5$$

Our goal is to find the maximum acceptable k for proper operation. This means finding k so the damping factor is no smaller than 0.5.

Assume the numerator $200k = 1 = 0$ dB—we'll find k later. Start plotting the asymptotes of G(s). The break frequencies are

$$\omega_1 = \frac{1}{0.2} = 5 \text{ rad/s}$$

$$\omega_2 = \frac{1}{0.05} = 20 \text{ rad/s}$$

$$\omega_3 = \frac{1}{0.01} = 100 \text{ rad/s}$$

See Fig. 1.29. Each break increases the downward slope by another 20-dB per decade.

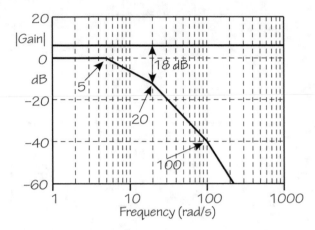

Figure 1.29: Determining k

Plot 1/H is in its proper place in the Bode diagram:

$$\frac{1}{H} = \frac{1}{0.5} = 2 \rightarrow 20\log_{10} 2 = 6 \text{ dB}$$

We know that 1/H must cross |G| before the second dive. This second break is at 100 radians/second. The question is how far can I shift upward the line for |G| to make 1/H cross at this second break?

I've marked the distance to raise |G| on the Bode plot in Fig. 1.29. The curve can be moved up as much as 18 dB.

Where I assumed the numerator is 1, we now see it can be 18 dB. So

$$\max |G| = 18 \text{ dB}$$
$$\text{so } 200k = 10^{18/20}$$
$$k = 0.0397$$

That's the largest k that doesn't produce a damping factor less than 0.5. If we increase the overall gain any more, we go below 0.5. We can choose a smaller gain, however, to get an even larger damping factor and even less oscillation.

1.7 SUMMARY

This has been a packed chapter! It covers two major topics, Bode plots and feedback, with two minor ones, dB and stability. But these all fit together to help us describe closed-loop control systems.

We use dB to compare signal levels, amplifier gains, and the like because it's a convenient yet logarithmic value. It is basic to Bode plots.

Bode plots are a quick way to lay out the asymptotes of a gain magnitude in dB versus the logarithm of frequency. While Bode was doing this as a way to get actual plots without a lot of calculation, we use this primarily for insight into system operation.

There are only three asymptotes in Bode plots as we have been using them: level, upward at 20 dB per decade, or downward at the same slope. These slopes can be integer multiples of 20 as well. Once we have the asymptotes, we can sketch the actual curve if we want to, but this isn't usually necessary.

Closed-loop control systems, sometimes called feedback systems, achieve more robust operation of a plant such as a motor or solenoid. By comparing the input signal with a report on the actual output, the system makes changes to make the output follow the input It does this even in the face of most perturbations.

For proper operation, the device or circuit that feeds back to the input a report on the output must be stable and precise. For the perfect system, the transfer function from input to output is the reciprocal of the feedback device's transfer function.

But unless the gain of the forward path is infinite, we can't have perfection. We use fractional error to measure how closely we come to this ideal system. Choosing an acceptable fractional error allows us to choose the necessary forward gain.

Controlling a plant with a closed-loop system improves two related characteristics of the plant's operation: response time and bandwidth. Our designs must consider appropriate amounts of underdamping, too. This can be determined from the Bode plot of $G(s)$ and $1/H$ by making sure that $1/H$ crosses $G(s)$ before the second break frequency of $G(s)$.

This discussion of closed-loop systems is very basic. It ignores most stability questions. It ignores systems with complex roots. So if you need to work extensively with such systems, do a lot more study!

Where do the parameters of the system elements come from? Where did we get the time constant for the positioner, for example? By appropriate observation of the system, we can find these parameters. That's the topic for Chapter 2.

FORMULAS AND EQUATIONS

1. Decibel as power and voltage ratios

$$\text{decibel} = 10\log_{10}\frac{P_{out}}{P_{in}} = 20\log_{10}\frac{V_{out}}{V_{in}}$$

2. Asymptotes

$|H(s)| = s$: Upward at 20 dB per decade, passing through $\omega = 1, dB = 0$

$|H(s)| = (1 + s/a)$: Flat at 0 dB to left of $\omega = a$; upward at 20 dB/decade to right

$|H(s)| = 1/(1 + s/a)$: Flat at 0 dB to left of $\omega = $ a; downward at 20 dB/decade to right

3. Bode's readjustments

3 dB inside corner

1 db inside at $\omega = $ half the corner frequency

1 dB inside at $\omega = $ twice the corner frequency

4. Classic closed-loop system

$$Y(s) = \frac{G(s)}{1 + G(s)H(s)}R(s)$$

5. Fractional error

$$E_{frac} = \frac{1}{G(s)H(s)}$$

6. Simplifying transfer functions

$$TF(s) = \frac{\dfrac{N(s)}{D(s)}}{1 + \dfrac{N(s)}{D(s)}H}$$

$$= \frac{N(s)}{D(s) + H \cdot N(s)}$$

7. Bandwidth where 1/H crosses |G| before second break

$$\omega_{new} = \omega_G\left(1 + |G_{DC}|H\right)$$

CHAPTER 2

Characterizing a System

Remember Chapter 1? Yes, right, the stuff on control systems. How did we happen to know the mathematical description of each device's transfer function Out(s)/In(s)? Transfer functions don't announce themselves on equipment labels! We did see, in Section 1.4, how one such function came from a scope plot of the positioner's step response. In this chapter we are going to look at how we can *characterize* a device to find that transfer function.

Why do we need this? Well, how do you design a control system that will control your linear device if you don't know its character? How do you determine the proper feedback to make the system do what you want it to do? How do you know the system is stable? How fast will it responds How robust it is in the face of varying conditions?

Here's an example. On an aircraft are flight controls that determine the motion of the plane. The aileron is a system that is controlled by the pilot. When the pilot moves a control, the aileron responds, altering the flight of the plane. The pilot would like some assurance that, when the control is moved, the aileron doesn't oscillate or start flapping but instead moves quickly and precisely to the desired position.

This aileron must move and get into the desired position in the face of varying forces on the aileron itself, changes in the plane's attitude, changes in temperature, changes in electrical voltage, and even changes in the pilot's attitude. The aileron must respond correctly every time.

You can't design the control system for this aileron without knowing something about either its time-domain (t) or its frequency-domain (s) characteristics. This chapter is about extracting those characteristics from a system.

We are going to restrict our look at characterization to linear systems, and to just first- and second-order systems. This should give you a start into the process that you might have to carry further later on.

2.1 TIME-DOMAIN RESPONSES

What are we characterizing? A linear system like what's shown in Fig. 2.1. This one has an input that is a function of time f(t) and an output x(t). These could be anything. The input could be a pressure in dynes and the output could be a position in furlongs, or degrees of rotation to produce current in microamperes, or position in inches to produce rotation in revolutions per parsec.

We will be in the frequency (s) domain since that's where we do much of control system design. The transfer function of a system is its output divided by its input. While my kooky systems are…well, kooky…they still can have transfer functions such as furlongs/dyne, microamperes/degree,

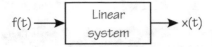

Figure 2.1: Linear system

or rpp/inch. But just to keep things simple, I'm going to assume that all our systems have voltage input and voltage output, so their transfer functions will generally be volts per volt.

2.1.1 FIRST-ORDER SYSTEMS

The time-domain response of a system is given by a differential equation that relates the output to the input as a function of time. The differential equation for the linear first-order system of Fig. 2.1 is

$$\tau \frac{dx(t)}{dt} + x(t) = kf(t)$$

This is a linear first-order differential equation with constant coefficients. It's about as simple as a differential equation can be. Also, it's in the form that "control people" like, with a "1" on the lowest-order term. ("Circuits people" tend to like having the "1" on the highest-order term.)

In this form, the d-c response is easy to get. If x is not varying with time, then $x = kf$ where f is constant as well.

Let's solve this equation via the Laplace transform and then take a look at the response to a particular input. First, transform the equation itself:

$$\tau\big(sX(s) - x(0)\big) + X(s) = kF(s)$$

Now set things up so that the initial value of x(t) is zero, which simplifies the equation:

$$\tau sX(s) + X(s) = kF(s)$$

which we solve to produce

$$X(s) = \frac{kF(s)}{\tau s + 1}$$

How does our system response to an input? We need to choose some input and see what the output does. But what input? There are two pretty standard inputs that we'll use to describe how a system responds. One, the unit impulse, gives us the *impulse response* of a system. The other input is the unit step, giving us the *step response* of a system. Either one of these yields the same information in the end, but I am going to stick with the step response.

So what does our example do with a step? We'll make f(t) a unit step (Fig. 2.2), transform that unit step via the Laplace transform, and see what happens:

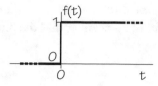

Figure 2.2: Unit step

$$f(t) = u(t) \quad \text{the unit step}$$

$$F(s) = \frac{1}{s}$$

$$X(s) = \frac{k}{s(\tau s + 1)}$$

The inverse Laplace transform produces the time-domain result, which is the familiar exponential:

$$x(t) = k\left(1 - e^{-t/\tau}\right)$$

Figure 2.3 shows this outcome.

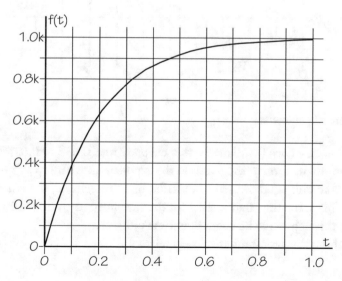

Figure 2.3: Exponential

There are several things to notice about this function:

- At $t = 0$ the output is zero. Even though the input jumped from 0 to 1, the output takes a while to get started; response is not instant.

- After a long time has passed, the output settles down to a *steady-state* value of k. So in the long run, our system has a gain of k, meaning that the output is eventually k times the input (which was 1).

- If we set $t = \tau$, then

$$x(\tau) = k\left(1 - e^{-\tau/\tau}\right) = k\left(1 - 0.368\right) = 0.632k$$

This is shown in Fig. 2.4.

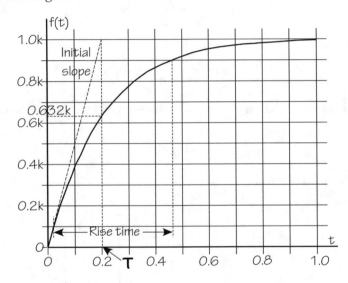

Figure 2.4: Exponential parameters

Note for future reference (i.e., later in this chapter) that $0.632 \approx 0.625 = 5/8$. So 5/8 is a close (1.1% error) approximation to the point on the exponential where $t = \tau$.

One other parameter is often used to describe this exponential, the *rise time*, which is also shown in Fig. 2.4. This is the time that it takes for the waveform to go from 10% to 90% of its final value. In the figure, this appears to be about 0.44 time units.

Rise time and the time constant are closely related:

$$t_{rise} = t_{90} - t_{10}$$
$$e^{-t_{90}/\tau} = 0.9, \quad e^{-t_{10}/\tau} = 0.1$$
$$-\frac{t_{90}}{\tau} = \ln 0.9, \quad -\frac{t_{10}}{\tau} = \ln 0.1$$
$$t_{rise} = -\tau\left(\ln 0.9 - \ln 0.1\right) = 2.197\tau$$

The rise time is about 2.2 times the time constant.

2.1.2   SECOND-ORDER SYSTEMS

Now we'll do the same analysis on a second-order system. The second-order differential equation is

$$\frac{1}{\omega_n^2}\frac{d^2 x(t)}{dt^2} + \frac{2\varsigma}{\omega_n}\frac{dx(t)}{dt} + x(t) = kf(t)$$

where two constants are important:

ω_n is the undamped natural frequency, which is the frequency at which the system will oscillate if there is no damping. We cannot observe this frequency in the real world, though, because damping is almost always present.

ζ is the damping ratio that determines how quickly the oscillation damps out, and even whether the oscillation appears at all. (The letter is *zeta*.)

Transforming and solving (assuming zero initial conditions) gives us X(s):

$$\left(\frac{s^2}{\omega_n^2} + \frac{2\varsigma}{\omega_n}s + 1\right)X(s) = kF(s)$$

$$X(s) = \frac{kF(s)}{\dfrac{s^2}{\omega_n^2} + \dfrac{2\varsigma}{\omega_n}s + 1}$$

The denominator of X(s) is called the *characteristic polynomial*. If we know the numbers for the two constants, we know the entire story about how the system responds. (In our first-order system, the characteristic polynomial is $\tau s + 1$.)

The second-order system isn't as simple as the first-order one is because...well, it's second order, a quadratic. It turns out that the roots of this polynomial are important when we want to describe the operation of our second-order system. So we need to look at these roots.

But first, let's go back to look at the root of the characteristic polynomial of our first-order system. The first-order equation is

$$\tau s + 1 = 0$$

which has just one root:

$$s = -\frac{1}{\tau}$$

This says that if we know the time-constant τ, we know the time response of this first-order system. All we need is the d-c constant k to finish the story.

The second-order system is characterized by the quadratic

$$\frac{s^2}{\omega_n^2} + \frac{2\varsigma}{\omega_n}s + 1 = 0$$

which has two roots,

$$s_{1,2} = -\varsigma\omega_n \pm \omega_n\sqrt{\varsigma^2 - 1}$$

It's the square root that makes this result interesting. The value of ζ determines how the system is going to respond, whether by oscillating like mad or only a little or not at all.

$$0 \leq \varsigma < 1 \quad \text{underdamped}$$
$$\varsigma = 1 \quad \text{critically damped}$$
$$\varsigma > 1 \quad \text{overdamped}$$

While all three of these outcomes are mathematically important, two of them aren't of much importance in control systems. Systems that are *overdamped* are sluggish in their response and we are therefore going to ignore $\zeta > 1$. Systems that are *critically damped* are impossible to make, because if the value of just one component drifts slightly with, say, temperature, ζ drifts away from 1 and the system is then either under or overdamped.

Now we focus on underdamping, where the square root is of a negative number. Let's use j, the square-root of −1, to get a different version of the two roots:

$$s_{1,2} = -\varsigma\omega_n \pm j\omega_n\sqrt{1 - \varsigma^2}$$

The underdamped system has an oscillation frequency that is defined by

$$\omega_d = \omega_n\sqrt{1 - \varsigma^2}$$

It is this frequency that we actually observe in a real system. This ω_d is called the *damped natural frequency.*

I haven't mentioned units of s and need to. Looking at the first-order response gives us a clue. The exponent of e must be unitless, so if t is in seconds, so is τ. Since the root of the first-order characteristic is $s = -1/\tau$, then s must have the unit of per-second. Because these are frequencies we label them *radians/second.*

The homogeneous solution (the solution with $f(t) = 0$) to our second-order differential equation is

$$x(t) = A_1 e^{s_1 t} + A_2 e^{s_2 t}$$

If we replace s_1 and s_2 by their values for this underdamped system,

$$x(t) = A_1 e^{-\varsigma\omega_n t} e^{j\omega_d t} + A_2 e^{-\varsigma\omega_n t} e^{-j\omega_d t}$$

Using Euler's formula gets this in sinusoidal terms:

$$x(t) = e^{-\varsigma \omega_n t} \left(B_1 \cos \omega_d t + j B_2 \sin \omega_d t \right)$$

Now let's apply as our input f(t) a unit step and see what the solution becomes:

$$x(t) = k \left(1 - \frac{1}{\sqrt{1 - \varsigma^2}} e^{-\varsigma \omega_n t} \cos \omega_d t \right)$$

Figure 2.5 is a typical underdamped second-order response to a unit step. There are lots of

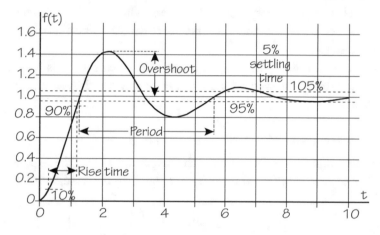

Figure 2.5: Second-order response

parameters to consider when we describe this second-order response. In the second-order equations, though, there are just three parameters, ς, ω_d, and ω_n. Let's look at some of the terms in Fig. 2.5:

- The most obvious in the diagram is the *overshoot*, which is how far beyond the final value the first peak goes. This value is generally expressed as a percent of the final value. In the drawing, the percent overshoot appears to be

$$\% \text{ overshoot} = 100 \frac{(1.42 - 1.00)}{1.00} = 42\%$$

- As in the first-order response, the *rise time* is the time the response takes to go from 10% to 90% of its final value. Here it appears to be about 0.9 seconds.

- The *period* is the time of one complete oscillation, here about 4.4 seconds.

- The *5% settling time* is the time it takes for the oscillation to settle to within 5% of its final value. In this waveform it looks like about 7.1 seconds.

We will be using these measurements from waveforms to characterize second-order systems in the next section.

2.2   CHARACTERIZING

Time to characterize! Yes, now! Here in public! In broad daylight, no less!

Characterizing really isn't that spectacular, and probably doesn't even border on exciting. What it means is finding the parameters of the differential equation that describes the system. For a first-order system, it means finding the time constant τ. For a second-order system, the undamped natural frequency ω_n and the damping ratio ζ

We do this by eliciting a step response of the system, which means we have to drive the system with a step input and see what happens. This "see" part will often be via sensors that display the outcome on an oscilloscope.

In the examples that follow, I'm going to assume that each system is linear, that it has been hit by a step input, and that we see the response as a trace on a carefully-adjusted scope so that the response fills the screen.

2.2.1   EXAMPLE I— 1st-ORDER SYSTEM

We've applied the step input to our system and the scope now shows the exponential response in Fig. 2.6. What's this second exponential that's upside down? In this case, we've used a square wave

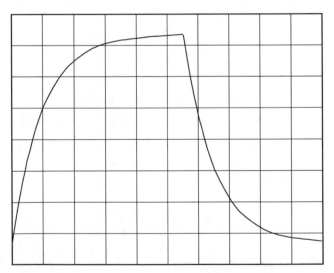

Figure 2.6: Example I: Scope before adjustments

to excite the system, which repeats itself—we are seeing the beginning of the second half of that repetition.

When making scope measurements, though, we need to pay attention to scope settings. In order to get the best precision, we need to use the full scope screen for our measurements, or as much of it as we can. This means adjusting both the vertical and horizontal gains on the scope to fill the screen with the "most measurable" trace.

Figure 2.7 is the result of this. I've made three adjustments:

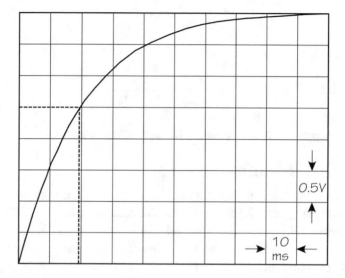

Figure 2.7: Example I: Scope with T measurement

- The trace now begins in the lower left corner.

- The flattening exponential "lands" along the top of the scope.

- The exponential is expanded horizontally so that only part of the waveform is shown—just enough to be sure of the "landing."

When we finish these adjustments, we read the screen calibrations: vertical 0.5V/□, horizontal 10 ms/□.

We'll calculate τ by measuring the time to the 1/e point on the trace. We learned in Section 2.1.1 that this point on a trace that is going up like this one does is at 0.632 of the full range and that 5/8 is a 1% approximation to that.

The trace starts in the lower left corner and fills the screen, "landing" at the top. So the trace covers eight squares. We'll read the time where the trace crosses the top of the fifth square. In Fig. 2.7, that looks to me like 1.9 squares from the left. At 10 ms/□, that's 19 ms, so for this system, $\tau = 19$ ms.

The trace covers eight squares, which is 4.0 volts. The input step was a unit step (i.e., 1 volt), so the d-c gain is 4.

The characteristic polynomial for our system is

$$19 \times 10^{-3} s + 1$$

and the differential equation becomes

$$19 \times 10^{-3} \frac{dx}{dt} + x = 4 f(t)$$

and the transfer function is

$$TF(s) = \frac{X(s)}{F(s)} = \frac{4}{19 \times 10^{-3} s + 1}$$

That's all we need to do to characterize a linear first-order system.

2.2.2   EXAMPLE II—2nd-ORDER SYSTEM

The second-order response to a unit step is more complex and so characterization is, too. We can't get the desired parameters (ζ and ω_n) directly. Also, the clarity of the "wiggles" of the second-order response determines which of two methods we use. We'll look at both here.

The scope picture in Fig. 2.8 is the result of a step of 1 volt, and the scope has been adjusted so the picture is "good," meaning that it fills most of the screen and there are only about two oscillations shown. After these adjustments, the vertical calibration is 0.5 V/□ and the horizontal calibration is 100 ms/□.

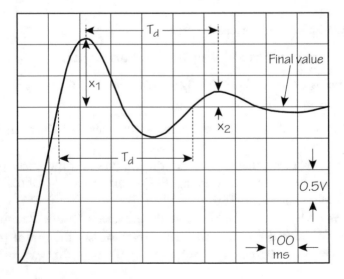

Figure 2.8: Example II: Using log decrement

This scope trace shows two clear upward "wiggles" and we can clearly read their heights above the final value of the trace. Since we can read these, we'll use the method of *logarithmic decrement*. (If the second one isn't clearly readable, we have to use the second method.)

From the scope (Fig. 2.8) we collect three pieces of data:

- Height of the first peak *above the final value:* $x_1 = 2.1$ squares

- Height of the second peak: $x_2 = 0.5$ squares

- Period T_d measured between peaks or between same-direction zero crossings: $T_d = 4.3$ squares

That's all the data we need to find the two parameters of this second-order system.

Start by finding the logarithmic decrement. Since this is a ratio, we don't need the actual scope calibration, just the numbers of squares:

$$\delta = \ln\left(\frac{x_1}{x_2}\right) = \ln\left(\frac{2.1}{0.5}\right) = 1.435$$

From this logarithmic decrement we get the damping ratio:

$$\varsigma = \frac{1}{\sqrt{1+\dfrac{4\pi^2}{\delta^2}}} = \frac{1}{\sqrt{1+\dfrac{4\pi^2}{1.435^2}}} = 0.223$$

The period that we measured is that of the *damped* response. The scope shows 4.3 squares at 100 ms/□, so the damped period is $T_d = 430$ ms. From this we get the *damped* natural frequency:

$$\omega_d = \frac{2\pi}{T_d} = \frac{2\pi}{430 \times 10^{-3}} = 14.6 \text{ rad/s}$$

Now we can find the *undamped* natural frequency:

$$\omega_n = \frac{\omega_d}{\sqrt{1-\varsigma^2}} = \frac{14.6}{\sqrt{1-0.223^2}} = 15.0 \text{ rad/s}$$

Finally, we can find the d-c constant k by noting that the input is a *unit* step, the final value is five squares up from the beginning. Since the scope calibration is 0.5 $V/\square$. $k = 2.5$.

Remember the characteristic polynomial and the differential equation for a second-order system?

$$\frac{s^2}{\omega_n^2} + \frac{2\varsigma}{\omega_n}s + 1$$

$$\frac{1}{\omega_n^2}\frac{d^2x(t)}{dt^2} + \frac{2\varsigma}{\omega_n}\frac{dx(t)}{dt} + x(t) = kf(t)$$

We have all of the numbers for these:

$$\frac{s^2}{15.0^2} + \frac{2(0.223)}{15.0}s + 1$$
$$= 4.44 \times 10^{-3}s^2 + 29.7 \times 10^{-3}s + 1$$
$$4.44 \times 10^{-3}\frac{d^2x(t)}{dt^2} + 29.7 \times 10^{-3}\frac{dx(t)}{dt} + x(t) = 2.5 f(t)$$

What happens if we can't reliably read the height of the second peak? If there's fairly heavy damping, this peak is often very small. Let's use the same waveform we just worked on, but let's assume we can't read the second peak. This will lead to the method of *fractional overshoot*.

The fraction in fractional overshoot is the ratio of the height of the first peak to the distance from the starting point to the final value. NOTE: This works only for a step input.

From the graph, $x_o = 5$ squares and $x_p = 2.1$ squares. The fractional overshoot is

$$OS = \frac{|x_p|}{|x_o|} = \frac{2.1}{5} = 0.42$$

This yields the damping ratio

$$\varsigma = \frac{1}{\sqrt{1 + \frac{\pi^2}{(\ln OS)^2}}} = \frac{1}{\sqrt{1 + \frac{\pi^2}{(\ln 0.42)^2}}} = 0.266$$

That's about what we got before. I'll admit I am carrying excess precision, especially considering that the data from the scope are good to only about two significant digits.

Sometimes we can't read a full period from the scope, either. In these cases, we can use the half period as shown in Fig. 2.9. On this trace, the half period $T_d/2 = 2.1$ squares, which is 210 ms. The damped natural frequency becomes

$$\omega_d = \frac{\pi}{(T_d/2)} = \frac{\pi}{210 \times 10^{-3}} = 15.0 \text{ rad/s}$$

Again, not too bad considering that we don't have great precision from the scope data. The characteristic polynomial and the differential equation for this system are about the same as before.

One additional quantity that is often used to describe a second-order system is the *settling time*. This is the time it takes for the oscillation to settle to within a certain percentage of its final value. Not uncommon are the 2% and the 1% settling times:

$$2\% \text{ settling time} = \frac{3.9}{\varsigma\omega_n}$$
$$1\% \text{ settling time} = \frac{4.6}{\varsigma\omega_n}$$

For our example (using the logarithmic decrement data), these are 1.2 s and 1.2 s.

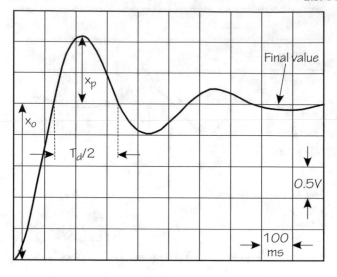

Figure 2.9: Example II: Using fractional overshoot

2.3   MORE EXAMPLE

Here are two more examples of characterizing a second-order system.

2.3.1   EXAMPLE III—LOGARITHMIC DECREMENT

The scope calibration on the time axis in Fig. 2.10 is 200 ms/□. Data from the screen give $x_1 = 2.15$, $x_2 = 0.45$, $T_d = 4.9 - 1.2 = 3.7$ squares, which is 740 ms.

The logarithmic decrement is

$$\delta = \ln\left(\frac{x_1}{x_2}\right) = \ln\left(\frac{2.15}{0.45}\right) = 1.564$$

which makes the damping ratio

$$\varsigma = \frac{1}{\sqrt{1 + \dfrac{4\pi^2}{\delta^2}}} = \frac{1}{\sqrt{1 + \dfrac{4\pi^2}{1.564^2}}} = 0.242$$

The damped natural frequency is

$$\omega_d = \frac{2\pi}{T_d} = \frac{2\pi}{740 \times 10^{-3}} = 8.49 \text{ rad/s}$$

and the undamped natural frequency is

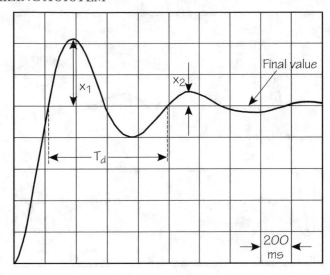

Figure 2.10: Example III

$$\omega_n = \frac{\omega_d}{\sqrt{1-\varsigma^2}} = \frac{8.49}{\sqrt{1-0.242^2}} = 8.75 \text{ rad/s}$$

The characteristic polynomial for Example I is

$$\frac{s^2}{8.75^2} + \frac{2(0.242)}{8.75}s + 1$$
$$= 13.1 \times 10^{-3} s^2 + 55.3 \times 10^{-3} s + 1$$

2.3.2   EXAMPLE IV—FRACTIONAL OVERSHOOT

In Fig. 2.11, the time axis calibration is 0.5 s/□. Data from this response have to be taken for fractional overshoot because the second peak is hard to measure. The results are $x_o = 5$, $x_p = 1.5$, and $T_d/2 = 2.0$ squares, which is 1.0 seconds.

Using the fractional overshoot

$$OS = \frac{|x_p|}{|x_o|} = \frac{1.5}{5} = 0.30$$

makes the damping ratio

$$\varsigma = \frac{1}{\sqrt{1+\dfrac{\pi^2}{(\ln OS)^2}}} = \frac{1}{\sqrt{1+\dfrac{\pi^2}{(\ln 0.30)^2}}} = 0.358$$

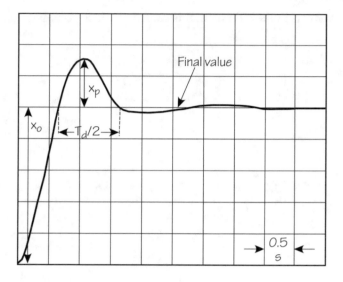

Figure 2.11: Example IV

The damped natural frequency is

$$\omega_d = \frac{\pi}{(T_d/2)} = \frac{\pi}{1.0} = 3.14 \text{ rad/s}$$

and the undamped natural frequency is

$$\omega_n = \frac{\omega_d}{\sqrt{1-\varsigma^2}} = \frac{3.14}{\sqrt{1-0.358^2}} = 3.36 \text{ rad/s}$$

2.4   SUMMARY

This chapter has been about "getting the numbers" from the response of a system. When we have a real, physical system and need to describe it mathematically, probably for designing a controller for the system, we need some clues as to how the system responds.

We've done just linear first- and second-order systems. The techniques here can be expanded to higher-order systems, which gets messy, and to non-linear systems, which gets even messier. Both of these are beyond what we are doing in this course.

I've chosen to use a step to excite these systems. We can also get the response by using a sinusoidal input, but that, too, is beyond our scope here.

Note that the outcome is the characteristic polynomial, which in the s domain completely describes the response of a linear system and therefore its transfer function. Since we generally design closed-loop control systems in the s domain, this characteristic is just what we need for the device we plan to control.

FORMULAS AND EQUATIONS

1. First-order linear system and s-domain solution.

$$\tau\frac{dx(t)}{dt} + x(t) = kf(t), \quad X(s) = \frac{kF(s)}{\tau s + 1}$$

$$t_{rise} = 2.2\tau$$

2. Second-order linear system and s-domain solution

$$\frac{1}{\omega_n^2}\frac{d^2x(t)}{dt^2} + \frac{2\varsigma}{\omega_n}\frac{dx(t)}{dt} + x(t) = kf(t), \quad X(s) = \frac{kF(s)}{\frac{s^2}{\omega_n^2} + \frac{2\varsigma}{\omega_n}s + 1}$$

3. Second-order system roots

$$s_{1,2} = -\varsigma\omega_n \pm \omega_n\sqrt{\varsigma^2 - 1}$$

$$0 \leq \varsigma < 1 \quad \text{underdamped}$$

$$\varsigma = 1 \quad \text{critically damped}$$

$$\varsigma > 1 \quad \text{overdamped}$$

4. Underdamped second-order system roots

$$s_{1,2} = -\varsigma\omega_n \pm j\omega_n\sqrt{1-\varsigma^2}, \quad \omega_d = \omega_n\sqrt{1-\varsigma^2}$$

5. Characterizing using logarithmic decrement

$$\delta = \ln\left(\frac{x_1}{x_2}\right), \quad \varsigma = \frac{1}{\sqrt{1 + \frac{4\pi^2}{\delta^2}}}$$

$$\omega_d = \frac{2\pi}{T_d}, \quad \omega_n = \frac{\omega_d}{\sqrt{1-\varsigma^2}}$$

6. Characterizing using fractional overshoot

$$OS = \frac{|x_p|}{|x_o|}, \quad \varsigma = \frac{1}{\sqrt{1 + \frac{\pi^2}{(\ln OS)^2}}}$$

7. Settling times

$$2\% \text{ settling time} = \frac{3.9}{\varsigma\omega_n}$$

$$1\% \text{ settling time} = \frac{4.6}{\varsigma\omega_n}$$

8. % Overshoot from damping ratio

$$\%OS = 100e^{-\varsigma\frac{\pi}{\sqrt{1-\varsigma^2}}}$$

CHAPTER 3

Instrumentation

Instrumentation—putting instruments into a system. Of course, there are all kinds of instruments, trombones, scalpels, even screwdrivers. Here, though—electrical instruments. Why electrical? Because we usually want a "reading" of a physical parameter some distance away from where the parameter is actually "happening."

A simple example of electrical instrumentation is the thermostat that controls the heat in your house. The device uses the thermal expansion of metal to detect changes in temperature. The movement of that metal is converted to an off-or-on electrical signal by a set of contacts. This electrical signal is delivered to the furnace to control the gas valve or the heat pump.

In this chapter I'm going to stick with just a couple of different physical quantities to be "read," strain and temperature. But before that, let's look at how we can use electrical information to determine the value of a physical quantity as precisely as possible.

Many instruments alter a voltage or a current associated with the parameters they sense. That alteration may be very small, sometimes only a few percent. So we are faced with reading a small change and we want to read this change as precisely as we can.

Let's presume that our gauge, whatever it is, has to operate on 10 volts. As the physical parameter is changed, that voltage is raised or lowered in a linear manner. Suppose the largest expected change is 4%, which means the voltage actually ranges from 9.6 to 10.4 volts. For my example, I'm going to read this using a digital voltmeter.

Now suppose I want to measure this parameter to a precision of one part in a thousand, 0.1%. At the maximum expected parameter value, the voltage is 10.4 V. I want to detect a 0.1% change, so I must be able to observe a change in the parameter from 4% to 4.004%. This means I need to be able to read a voltage of 10.4004 volts. To do this, I must have a voltmeter that is precise to about six significant digits. Not unheard of, but expensive.

Let's get clever and build a circuit that somehow precisely subtracts that "base voltage" of 10 volts from the gauge's signal. The circuit that does this subtraction must be precise and stable, but it turns out this is not hard to do.

After I precisely subtract 10 volts, the range of the change of voltage from the gauge is −0.4 to +0.4 volts. At the +0.4 V end, 0.1% means at most 0.4004 VV, a precision of just four significant digits. That's not as hard to do as the original six. In the next section we'll see how a Wheatstone bridge does this "subtracting."

3.1   STRAIN GAUGE

The resistance strain gauge is simple, precise, and inexpensive. It measures strain, usually stated in percent elongation. It works by stretching or compressing a conductor whose dimensions are carefully controlled. Figure 3.1 shows how the conductor is laid out on the substrate. Figure 3.2 shows the size of a typical gauge.

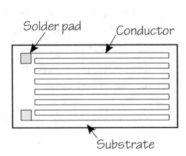

Figure 3.1: Strain gauge

Figure 3.2: Strain gauge

 The electrical parameter that changes with strain is the resistance of the conductor. The change in resistance is directly proportional to the strain the gauge undergoes. To use the gauge, you glue it to the surface to be tested and run wires to the metering system. Figure 3.3 shows two gauges glued to an aluminum bar.

 Gauges are not expensive. Omega sells standard strain gauges for around $50 for a pack of ten, for example.

 The resistance of a conductor depends on three parameters, conductivity (a property of the metal the conductor is made from), length, and cross-sectional area:

$$R = \frac{L}{\sigma A}$$

Figure 3.3: Strain gauge attached to beam

It is the length parameter that we are most concerned with. The change in length is proportional to strain and the change in resistance is proportional to length. Hence, the change in resistance is proportional to strain.

There's another effect operating here, though, that we at least have to be aware of. As the conductor lengthens, its cross-sectional area must decrease because there is a constant volume of metal. Increasing the length increases the resistance, but it also decreases the area, which also increases the resistance. This effect is a second-order effect, though, and for many measurements we can ignore it. A check of the manufacturer's literature will tell you how much this decrease in area affects the measurement.

But wait! We aren't done with "effects" yet. The strain gauge also has what is called *transverse sensitivity.* This is a measure of how much the gauge is affected by a strain at right angles to the desired strain and the "long way" position of the gauge. Note in Fig. 3.1 the little sections of conductor that run across the gauge where the conductor turns a corner. These little sections are sensitive to transverse strain, or strain across the beam. The package in Fig. 3.2 shows a transverse sensitivity of $(0.4\pm0.2)\%$, which says that the resistance of the gauge is affected by this transverse strain only to at most 0.6% of the gauge's resistance.

For a typical strain gauge, the strain, expressed in terms of the change in resistance is

$$\varepsilon \approx \frac{\Delta R}{R}$$

where R is the unstrained resistance of the gauge.

For the gauge itself, the change in resistance is

$$\Delta R = G\varepsilon R$$

where G is a gauge constant, ε is the strain, and R is the gauge resistance (unstrained). Gauges are commonly made with standard values of R: 120, 240, 350, and 1,000 ohms. A common gauge factor is about 2. Note in Fig. 3.2 that the gauge factor G is 2.090±0.5% and the gauge resistance R is 120.0±0.3% Ω

Gauges are also affected by temperature, which should be taken into account for very precise measurements. Omega, for example, makes gauges whose temperature coefficient is matched either to steel or to aluminum.

3.1.1   VOLTAGE DIVIDER

I want to get from a gauge a voltage that varies with strain. The simplest circuit I can think of is the voltage divider (Fig. 3.4), where R_2 is the gauge. I'll choose a second resistance R_1 to be 120 ohms,

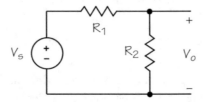

Figure 3.4: Strain gauge in voltage divider

equal to R_2, the gauge resistance. This makes $V_o = \frac{1}{2}V_s$ for zero strain.

In fact, if I want a resistor for R_1 that is as precise as the gauge resistance, why not just use a second gauge? This "other gauge" is placed so that it will not undergo any strain and hence remain 120 ohms. Moreover, this "other gauge" will be affected by temperature in the same way as the actual gauge and to some extent balance out the effect of temperature change. (Gauges are pretty inexpensive.)

Now write the equation for the voltage divider:

$$V_o = V_s \frac{R_2}{R_1 + R_2}$$

Replace the gauge resistance R_2 with its total resistance that includes ΔR:

$$R_2 = R_g + \Delta R = R_g + G\varepsilon R_g = R_g(1 + G\varepsilon)$$
$$R_1 = R_g$$
$$V_o = V_s \frac{R_g(1 + G\varepsilon)}{R_g + R_g(1 + G\varepsilon)} = V_s \frac{1 + G\varepsilon}{2 + G\varepsilon}$$

Suppose that the gauge factor G is 2.00 and the supply voltage $V_s = 5.0000$ V. For zero strain, $V_o = 2.5000$. For 1% strain:

$$V_{o1} = 5\frac{1+0.02}{2+0.02} = 2.524752 \text{ V}$$

Suppose we want to read the strain to a precision of 0.001. If the strain is 1.001%, the voltage V_o will be

$$V_{o1o01} = 5\frac{1+0.02002}{2+0.02002} = 2.524777 \text{ V}$$

Precision of 0.001 is a voltage difference of only 0.000025 V, which is 25 microvolts. The voltmeter must be precise to six digits after the point, seven digits total.

3.1.2   WHEATSTONE BRIDGE

My voltage-divider example shows that it would be very nice if we could somehow subtract out the zero-strain voltage. In the example, the zero-strain voltage was 2.50000 V, while the voltage for 1% strain was 2.524752 V. If we could somehow subtract 2.5 volts, the 1% strain would read 0.024752 volts. We could amplify this by, say, a factor of 100 so the voltmeter would read 2.4752 volts. Now a precision of 0.001 requires only that the voltage be read ± 0.0025 volts, which is 2.5 millivolts. That requires only four digits after the point in our voltmeter.

The Wheatstone bridge was invented in 1833 by … no, not Sir Charles Wheatstone (1802–1875) but by Samuel Christie. It implements a concept sometimes called *differential measurement*. This circuit balances out the zero-strain voltage and gives a reading of just the voltage associated with the strain itself. The bridge in Fig. 3.5 has two strain gauges and two resistors in it. The two strain gauges are in the upper arms; the two resistors are in the lower arms.

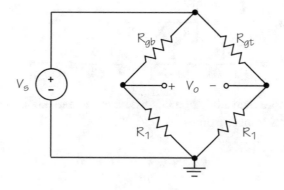

Figure 3.5: Strain-gauge bridge—two arms

A clever part of this circuit is in the placement of the strain gauges on the beam being stressed. One gauge, on the upper left, is attached to the bottom of the beam; the other gauge, on the upper right, is attached to the top of the beam. If the beam is being deflected downward, the top gauge (upper right arm) sees tension and positive strain while the bottom gauge (upper left arm) sees compression and negative strain. This doubles the sensitivity of the measurement.

The two resistors in the lower arms of the bridge are precision resistors whose resistance is identical to that of the strain gauges. If for example we use 120-Ω gauges, the bottom resistors will be 120 Ω also.

Analysis of this bridge circuit is done using the voltage-divider relationship. But remember that a voltage divider is a valid method only when the divider doesn't "leak," meaning that there is no current being drawn from the divider itself. That's the situation here because we are going to put a voltmeter between the middle terminals of the bridge. That meter will draw only a very tiny current from the terminals (microamperes) and will barely affect the voltage dividers. Moreover, we will find that when the bridge is exactly balanced, i.e., all four resistances are exactly equal, the voltage between the terminals will be zero.

Let's start the analysis. The voltages referenced to ground at the plus and the minus terminals are

$$V_+ = V_s \frac{R_1}{R_1 + R_{gb}}, \quad V_- = V_s \frac{R_1}{R_1 + R_{gt}}$$

Now apply the strain gauge equations

$$R_{gb} = R_g - G\varepsilon R_g, \quad R_{gt} = R_g + G\varepsilon R_g$$
$$R_1 = R_g$$

The two signs account for the fact that the bottom gauge is compressed when the beam is pressed downward, while the upper gauge is stretched.

The voltmeter reads the difference:

$$V_o = V_+ - V_- = V_s \left[\frac{R_g}{R_g + R_g - G\varepsilon R_g} - \frac{R_g}{R_g + R_g + G\varepsilon R_g} \right]$$
$$= V_s \left[\frac{1}{2 - G\varepsilon} - \frac{1}{2 + G\varepsilon} \right] = V_s \frac{2G\varepsilon}{4 - G^2\varepsilon^2}$$

If the strain is no more than 0.04, then $G^2\varepsilon^2$ is no more than 0.0064 for a gauge factor $G = 2$. If $G^2\varepsilon^2 \ll 4$, the voltage formula becomes

$$\text{for } \varepsilon \ll 1, \quad V_o = \frac{V_s}{2} G\varepsilon$$

When the strain is zero, so is V_o. For $V_s = 5.0$ V, $G = 2$ and $\varepsilon = 1\%$, the bridge voltage $V_o = 0.05$ V, a voltage that within the range of any decent digital voltmeter.

3.1.3   IMPROVED BRIDGE

In the previous example, the two gauges were in the upper arms of the bridge. One was attached to the tension side of the beam; the other, to the compression side. We could use for the two lower arms two more gauges but *unattached* to the beam. This isn't a bad idea for several reasons: gauges are cheap, they provide exactly the right resistance, and they give all four arms the same temperature coefficient.

Hmmmm, since we have two unstressed gauges, could they be attached to the beam as well? Yup! We'll attach all four gauges to the beam, two on the top and two on the bottom (Fig. 3.6). The

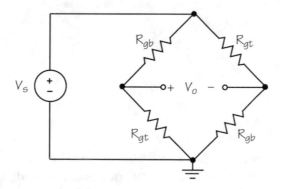

Figure 3.6: Strain-gauge bridge—four arms

gauges on the same side of the beam are in diagonally opposite arms of the bridge. The photo in Fig. 3.3 shows two gauges on top of a beam. The other two are not visible underneath.

The analysis of the circuit of Fig. 3.6 proceeds as before.

$$V_o = V_s \left[\frac{R_{gt}}{R_{gb} + R_{gt}} - \frac{R_{gb}}{R_{gb} + R_{gt}} \right]$$

$$R_{gt} = R_g + G\varepsilon R_g, \quad R_{gb} = R_g - G\varepsilon R_g$$

$$R_{gt} + R_{gb} = 2R_g$$

$$V_o = V_s \left[\frac{R_g(1 + G\varepsilon)}{2R_g} - \frac{R_g(1 - G\varepsilon)}{2R_g} \right] = \frac{V_s}{2} 2G\varepsilon$$

$$V_o = V_s G\varepsilon$$

The outcome is a measurement that is twice as sensitive as the two-arm bridge.

There's one more adjustment that we need to make, though. I have been assuming all through this bridge stuff that the four arms of the bridge are, in their unstressed state, exactly equal. I need this to be true so that the bridge voltage V_o is exactly zero for zero stress. But alas, that's not always going to be 100% true, so we need a way to make a minor adjustment to the bridge. Figure 3.7 shows one way of doing this.

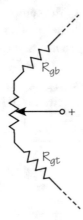

Figure 3.7: Bridge adjustment

The added resistance between the two left arms of the bridge is an adjustable resistor. Once the gauges and the bridge are set up and before any gauges are stressed, the adjustable resistance, called a "pot," (short for *potentiometer*), is adjusted by moving the tap (the arrow in the drawing) up or down until the voltmeter reads exactly zero. Then the bridge is ready to read strain via the voltmeter readings.

3.1.4   EXAMPLE I–2-ARM STRAIN-GAUGE BRIDGE

Two strain gauges ($R = 120\Omega$, $G = 2.090$) are set up in the two-arm bridge arrangement of Fig. 3.5. The bridge supply is $V_s = 5.0$ V. The voltmeter reads $V_o = 86.213$ mV. Find the strain in percent.

For the two-arm bridge, the voltage relationship is

$$V_o = \frac{V_s}{2} G \varepsilon$$

For the data given,

$$86.213 \times 10^{-3} = \frac{5}{2} 2.090 \varepsilon$$

so

$$\varepsilon = 1.65\%$$

3.1.5   EXAMPLE II–4-ARM STRAIN-GAUGE BRIDGE

Four gauges ($R = 120\ \Omega$, $G = 2.090$) are set up in a four-arm bridge arrangement of Fig. 3.6. The supply voltage is 5.0 V and the voltmeter reads $V_o = 86.213$ mV. What is the indicated strain?

For the four-arm bridge,

$$V_o = V_s G \varepsilon$$

The result is

$$86.213 \times 10^{-3} = (5)(2.090)\varepsilon$$
$$\varepsilon = 0.825\%$$

Note that this four-arm bridge is twice as sensitive as the two-arm bridge.

3.2   RESISTANCE TEMPERATURE DETECTOR

There are lots of common ways to measure temperature, or I suppose I really should say, to *sense* temperature. Different sensors have different properties:

Thermometer: Changing volume, very common, inexpensive, hard to "electrify" for a control system

Thermocouple: Self-generated voltage, very inexpensive, covers wide range of temperature, small, but requires reference junction

Integrated circuit: Transistor-like element varies current, inexpensive, generally linear with temperature, rugged, sensitive

Thermistor: Varying resistance, rather fragile, narrow range

Resistance temperature detector: Varying resistance, precision sensor, stable and accurate, expensive

In this section we are going to examine the resistance temperature detector because it is a rugged sensor designed to be used in a variety of situations. We apply the Wheatstone bridge as we did with strain gauges. Later we'll look at thermocouples because they are used in a manner that is very different from the resistance temperature detector.

3.2.1   RTD ITSELF

The RTD (one gets tired of saying the whole name!) is a standardized resistance with a very predictable and repeatable temperature characteristic. One popular type is a platinum resistance with a 0°C resistance of 100 ohms. (See Fig. 3.8.) The resistor itself often has four wires connected rather than two.

Figure 3.8: RTD

The sensor can be obtained in different "packages," depending on the application. Fig. 3.9 shows a model with the resistive element encased in a stainless steel tube. The resistance element is at the tip of the tube.

Figure 3.9: Resistance temperature detector

The resistance of the gauge R_g varies in a fairly friendly way with temperature. The actual relationship is a power series, the first terms of which are the Callendar–VanDusen equation:

$$R_g = R_o\left(1 + 3.9083 \times 10^{-3}T - 5.775 \times 10^{-7}T^2\right)$$
$$0 \leq T \leq 850°C$$

for $T < 0°C$, the equation is different (different coefficients and a T^3 term).

We often use an "official" linear equation with a *DIN-standard* coefficient, now the official International Electrotechnical Commission (IEC) standard:

$$R_g = 100(1 + 0.00385T) \text{ for T in } °C$$

The error between this linear equation and the polynomial is 0.007% at 100°C, 0.648% at 200 °C, and 2.805% at 400°C. We'll use the linear standard in the examples.

3.2.2   2-WIRE RTD BRIDGE

Using the RTD in an instrumentation system is a lot like using a strain gauge. We'll use a bridge just as we did with the strain gauge. But there's a major difference—the RTD is expensive ($100) so there will be just one of them in the bridge. (Besides, think about how one might deploy four RTDs like we did strain gauges!)

Figure 3.10 is the "2-wire" bridge circuit; i.e., two wires connecting the RTD. The three resistor arms are chosen to be exactly 100 Ω. We could also include a pot in the left corner just as we did for adjusting the strain gauge bridge (Fig. 3.7), but I won't do that here.

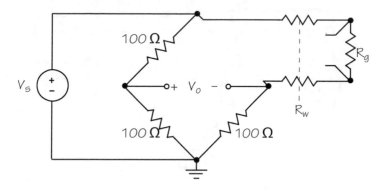

Figure 3.10: RTD 2-wire bridge

The complication is that the RTD's connecting wires have resistance that cannot be ignored. For example, if the two wires add just one ohm total, their effect is to make the indicated temperature appear to be 2.6°C higher.

Analysis is the same as with the strain gauge. There are two voltage dividers and V_o is the difference between the voltages of the two dividers. The wire resistance R_w is present twice in the upper right arm. The voltmeter draws essentially no current.

The two voltage divider equations are

$$V_{o+} = \frac{1}{2} V_s, \quad V_{o-} = V_s \frac{100}{100 + R_g + 2R_w}$$

Combining these and simplifying gives us the 2-wire bridge equation:

$$V_o = V_{o+} - V_{o-} = V_s \frac{0.385T + 2R_w}{400 + 0.77T + 4R_w}$$

To be able to make a calculation of temperature T from a reading of the voltage V_o, we need to know the wire resistance R_w, but this can generally be measured after installation of the sensor.

3.2.3   3-WIRE RTD BRIDGE

Remember the two extra wires on the RTD in the drawing of Fig. 3.8? Those two extra wires come in handy for both the three-wire bridge and the four-wire ohmmeter.

Figure 3.11 shows the "three-wire" bridge arrangement; i.e., three wires connecting the RTD. Notice that the V_{o-} terminal is connected to one of the extra wires, not to the right-hand corner of the bridge. Now consider why.

The voltmeter draws essentially zero current, so the current through the wire resistance of this third wire is zero—or very close to it. Hence, there is no voltage drop along that wire. While this

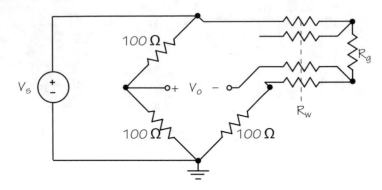

Figure 3.11: RTD 3-wire bridge

sounds like it might balance out all wire resistance, it doesn't, as the following equations show. I'm going to get V_o in terms of T in the same way I did it for the two-wire bridge. This time, there is an R_w in both the upper and lower arms on the right.

$$V_{o+} = \frac{1}{2}V_s, \quad V_{o-} = V_s \frac{100 + R_w}{100 + R_g + 2R_w}$$

$$V_o = V_{o+} - V_{o-} = V_s \frac{0.385T}{400 + 0.77T + 4R_w}$$

The difference between this and the two-wire equation doesn't seem to be much. Here, the numerator has no $2R_w$ term. This has the effect of cancelling out some but not all of the effects of the wire resistance.

Why didn't we have this wire-resistance problem with strain gauges? Because in the four-arm strain-gauge bridge, all the wire resistances balance out.

3.2.4   COMPARISON OF TWO BRIDGES

Let's compare the two bridge methods for reading an RTD. For something to work with, I'll use a supply voltage $V_s = 10$ V and assume the temperature T = 20 °C.

I'll start with no wire resistance, so putting $R_w = 0$ in each of the equations gives the "perfect" result:

$$V_{20} = 10 \frac{(0.385)(20)}{400 + (0.77)(20)} = 185.36 \text{ mV}$$

Now make the wire resistance $R_w = 0.1 \,\Omega$. For the two-wire bridge:

$$V_{20} = 10 \frac{(0.385)(20) + (2)(0.1)}{400 + (0.77)(20) + (4)(0.1)} = 190.00 \text{ mV}$$

For the three-wire bridge:

$$V_{20} = 10\frac{(0.385)(20)}{400 + (0.77)(20) + (4)(0.1)} = 185.19 \text{ mV}$$

The result is that the two-wire bridge gives a result 2.50% larger than the "perfect" value, while the three-wire bridge gives a result that is 0.09% less.

3.2.5   4-WIRE OHMMETER

Would you believe there's still one more configuration—and it completely eliminates the wire resistances? So you might just wonder why we even bothered with the bridge circuits. The reason is that this final method tends to be more expensive.

The circuit for the four-wire ohmmeter is Fig. 3.12. It works as follows:

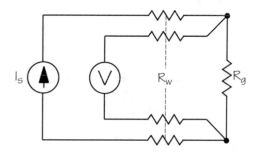

Figure 3.12: RTD 4-wire ohmmeter

- The current source I_s is precise, which means that its value is known and constant.

- The voltmeter is connected through the other pair of wires. It draws essentially no current through those wire resistances, so the voltage it measures is the actual voltage across the gauge resistance R_g.

- The current I_s flows through the RTD. None of it goes to the voltmeter.

- Since all of the current I_s flows through the gauge resistance R_g and the voltmeter truly indicates the voltage across R_g, the result is

$$V_m = I_s R_g = (I_s)(100)(1 + 0.00385T)$$

From this formula we can get T directly.

3.2.6 EMPLOYING AN RTD

You can't just stick an RTD into a tank or a pipe or a duct and expect perfect performance. So what should you think about? Several things come to mind:

- Immersion in the fluid so that the RTD is truly "seeing" the temperature and isn't in some kind of pocket

- Wiring that is as immune as possible from outside electrical interference (Chapter 5)

- Current through the RTD chosen so that the RTD is not self-heating, meaning that the current is not heating the resistance element enough to give a false temperature reading

Consider the last one for a moment. In my comparison of the two bridge methods (Sec. 3.2.4), I chose a bridge supply voltage $V_s = 10$ V. In both bridge circuits (Figs. 3.10 and 3.11), this voltage drives both voltage dividers. Ignore wire resistance for the moment. The current through each bridge arm is

$$i_{arm} = \frac{10}{100 + 100} = 50 \text{ mA}$$

This current heats the RTD by plain ordinary i^2R heating:

$$P_{RTD} = 100i_{arm}^2 = 0.25 \text{ W}$$

The hard question is, is this too much heat? The answer is, it depends! If the RTD is in good thermal contact with a large thermal mass that it's sensing, maybe this is OK. This is simply something that you need to watch out for.

Notice that if I had chosen a bridge supply of just one volt, the current would be one-tenth as large and the heating would be one-hundredth as large—just 2.5 mW.

3.2.7 EXAMPLE III–COMPARING RTD CONNECTIONS

An RTD is connected using two different bridge connections, one at a time. The wire resistance for both is $R_w = 0.2 \, \Omega$. In keeping with what I just said about self-heating, I'm setting the power supply to $V_s = 1$ V. In each case, the voltage reading is 80.000 mV. What temperature is being sensed?

The equation for the two-wire bridge with the numbers filled in is

$$0.08 = 1\frac{0.385T + (2)(0.2)}{400 + 0.77T + (4)(0.2)}$$

which when solved for temperature T gives

$$T = 97.910°C$$

Doing the same things with the three-wire bridge gives a slightly different result:

$$0.08 = 1\frac{0.385T}{400 + 0.77T + (4)(0.2)}$$
$$T = 99.147°C$$

Which one is right? The simple answer is, Yes! Yes, they are both right; no, they aren't. The meter is reading a particular voltage, namely, 80 mV. The outcome is affected by the choice of bridge.

What happens when we think a little differently? Suppose the actual temperature has been correctly sensed by the two-wire bridge. If we change to a three-wire bridge, what voltage does our meter read:

$$V_o = 1\frac{(0.385)(97.910)}{400 + (0.77)(97.910) + (4)(0.2)}$$
$$V_o = 79.160 \text{ mV}$$

The wiring makes a difference!

3.2.8   EXAMPLE IV–RTD OHMMETER

Suppose the actual temperature is the value we just found using the two-wire bridge: $T = 97.910°C$. What voltage do we get from the four-wire ohmmeter measurement?

I'm going to choose a current $I_s = 5$ mA because that's what the current was in both of the previous examples (1 volt divided by 200 ohms). Here is the calculation:

$$V_m = I_s R_g = (5 \times 10^{-3})(100)(1 + 0.00385 \times 97.910)$$
$$V_m = 688.48 \text{ mV}$$

This method gives a much higher voltage to observe on the voltmeter for the same amount of self-heating.

3.3   THERMOCOUPLE

Thermocouples are great! They are so-o-o-o simple. Just twist together a couple of wires made of different metals and you'll get a voltage that is very predictably related to the temperature of the junction. Golly, what could be neater?

But wait! If you twist together two wires of differing metals, you'll have to join two wires of differing metals somewhere else in the circuit as well. Otherwise, you won't have a closed loop. No loop, no circuit. Thermocouples are simple but you can't build a single junction into a circuit.

3.3.1   SEEBECK EFFECT

Thomas Johann Seebeck (1770–1831) discovered the effect that bears his name. It says that if you make a junction of two dissimilar metals and heat the junction, you'll get a voltage. This isn't a big voltage, typically something in the neighborhood of 40 μV/°C, so even 1,000°C will yield only 40 mV. But there's a voltage nevertheless.

Figure 3.13 shows this simple arrangement. The voltage produced depends on what metals you choose for the wires and it depends on the temperature. The relationship between voltage and temperature isn't linear, either. The polynomial relationships between voltage and temperatures for various combinations of metals are well documented.

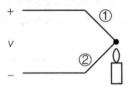

Figure 3.13: Seebeck effect

But as I said, you can't have a junction of two dissimilar metals without a second junction. Figure 3.14 shows one arrangement with two different metals, a voltmeter with copper terminals, and a ice-water bath for a constant 0°C for the second junction. The strange names are specialized alloys made for thermocouples. (See Fig. 3.18.)

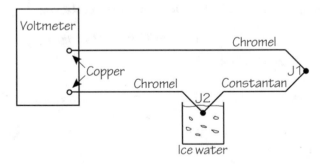

Figure 3.14: Thermocouple system I

The second junction J2 is called the *reference junction* because it provides the base temperature to which the temperature of the junction J1 is referenced. The measured voltage is therefore caused by the difference in temperature between the heated junction and the 0°C reference junction.

This reference could be at any fixed, stable temperature, not just 0°C. If we choose some other reference, then the voltage represents the temperature difference between the two. If, for example, we have a well-regulated oven at exactly 27°C, then the voltage is related to T − 27.

3.3.2 OTHER JUNCTION ARRANGEMENTS

Ice water is pretty inconvenient as a permanent feature of a temperature-measuring system, so we often use a block of material that can be kept by some means at a constant temperature. This block is called an *isothermal block* because its temperature is the same all over the block.

Figure 3.15 uses such a block to keep the two copper terminals of the voltmeter cable at the same temperature. The purposes of this is to control any small temperature difference between the voltmeter cable connections. The second junction J2 is still in the ice-water bath.

Figure 3.16 expands the block to include the second junction. The thinking here is that the block is being held at a constant temperature for the cable connections, so why not use it for the reference as well.

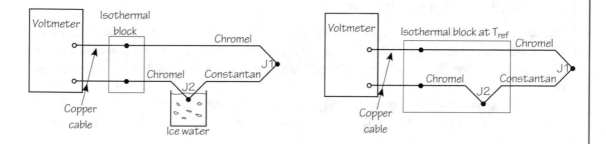

Figure 3.15: Thermocouple system II

Figure 3.16: Thermocouple system III

Figure 3.17 goes to the ultimate and just lets the voltmeter cable terminate both of the metals of the thermocouple. In this arrangement and in the previous one, the block is providing the reference temperature.

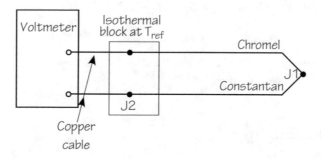

Figure 3.17: Thermocouple system IV

Any way you do it, you still have to maintain a second junction at a steady reference temperature. Now, what can go wrong with this simple arrangement? Here are a few problems:

- Bad connections, which interfere with the current from the very small junction voltages

- Wrong extension cables that create extra but uncontrolled junctions (there are cables designed for use in thermocouple circuits)

- Lead resistance, which means the meter current must be kept very low

- Junction diffusion, whereby at higher temperatures, one metal diffuses into the other, decalibrating the junction

- Noise insertion on the cable, interfering with the small junction voltages (Example III in Chapter 5).

3.3.3   THERMOCOUPLE TYPES

There are many different metal combinations that can be used to build thermocouples. All of them take two dissimilar metals and make two junctions, one to sense the temperature and the other to be the reference. Over the years, various strange-sounding alloys have been developed for thermocouples.

I'm going to show four fairly common types of thermocouples. The table of Fig. 3.18 lists the characteristics of Types E, J, K, and T. I've ignored those that use platinum alloys such as Types B, R, and S.

Type	Positive metal	Negative metal	Approx. sens'ity	Temp. range	Notes
E	Chromel	Constantan	68 μV/°C	−100→1000 °C	Non-magnetic
J	Iron	Constantan	55 μV/°C	−40→750°C	Inexpensive
K	Chromel	Alumel	41 μV/°C	−200→1350 °C	Magnetic effects
T	Copper	Constantan	43 μV/°C	−200→350 °C	One copper lead

Alumel™: 95% Ni, 2% Mn, 2% Al, 1% Si
Chromel: 90% Ni, 10% Cr
Constantan: 55% Cu, 45% Ni

Figure 3.18: Some thermocouple types

Each of these types is described by a power series, the coefficients of which are well documented. All the polynomials are of the form

$$\Delta T = \sum_{i=0}^{n} a_i V_s^i$$

where the coefficients are those of Fig. 3.19. The voltage V_s is the observed Seebeck voltage in **micro**volts, and ΔT is the temperature difference between the junctions in Celsius.

	Type E	Type J	Type K	Type T
a0	0.000 000 0	0.000 000	0.000 000	0.000 000 ....
a1	$1.705\ 703\ 5 \times 10^{-2}$	$1.978\ 425 \times 10^{-2}$	$2.508\ 355 \times 10^{-2}$	$2.592\ 800 \times 10^{-2}$
a2	$-2.330\ 175\ 9 \times 10^{-7}$	$-2.001\ 204 \times 10^{-7}$	$7.860\ 106 \times 10^{-8}$	$-7.602\ 961 \times 10^{-7}$
a3	$6.543\ 558\ 5 \times 10^{-12}$	$1.036\ 969 \times 10^{-11}$	$-2.503\ 131 \times 10^{-10}$	$4.637\ 791 \times 10^{-11}$
a4	$-7.356\ 274\ 9 \times 10^{-17}$	$-2.549\ 687 \times 10^{-16}$	$8.315\ 270 \times 10^{-14}$	$-2.165\ 394 \times 10^{-15}$
a5	$-1.789\ 600\ 1 \times 10^{-21}$	$3.585\ 153 \times 10^{-21}$	$-1.228\ 034 \times 10^{-17}$	$6.048\ 144 \times 10^{-20}$
a6	$8.403\ 616\ 5 \times 10^{-26}$	$-5.344\ 285 \times 10^{-26}$	$9.804\ 036 \times 10^{-22}$	$-7.293\ 422 \times 10^{-25}$
a7	$-1.373\ 587\ 9 \times 10^{-30}$	$5.099\ 890 \times 10^{-31}$	$-4.413\ 030 \times 10^{-26}$	
a8	$1.062\ 982\ 3 \times 10^{-35}$		$1.057\ 734 \times 10^{-30}$	
a9	$-3.244\ 708\ 7 \times 10^{-41}$		$-1.052\ 755 \times 10^{-35}$	
Range	$0 \to 1000°C$	$0 \to 760°C$	$0 \to 500°C$	$0 \to 400°C$
Error	$+0.02 \to -0.02°C$	$+0.04 \to -0.04°C$	$+0.04 \to -0.05°C$	$+0.03 \to -0.03°C$

Figure 3.19: NIST thermocouple polynomial coefficients

If you need to employ thermocouples, a most important facet of your job is to pay close attention to metals and junctions, especially unwanted or uncontrolled ones.

3.4   MORE EXAMPLE

3.4.1   EXAMPLE V–BROKEN STRAIN-GAUGE BRIDGE

Four strain gauges ($R = 120\ \Omega$, $G = 2.090$) are connected in a four-arm bridge (Fig. 3.6). The power supply $V_s = 5.0$ V. The voltmeter is presently reading $V_o = 0.15675$ V.

The gauge in the upper right arm of the bridge fails open (see Fig. 3.20). In other words, the gauge is replaced by an open circuit. What does the voltmeter read now?

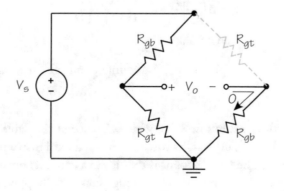

Figure 3.20: Bridge with broken arm

Remember that the voltmeter draws a very small current, small enough that the current doesn't affect the bridge. The meter's minus lead is still connected to the right corner of the bridge. Now, though, that right corner has only one gauge connected to it, the gauge in the lower right arm. The other gauge has become an open circuit.

The voltmeter's current is so small that there is negligible voltage drop across that lonely gauge on the lower right. Therefore, the voltage of the minus terminal is essentially that of the ground reference. So we can write

$$V_- = 0$$

and therefore the voltmeter reads

$$V_o = V_+ - V_- = V_+$$

To finish this, we need the voltage V_+, which is not $V_S/2$ because the gauges are stressed. To get this, we need to know the gauge resistances. To get the gauge resistances, we need the strain that was being experienced. For the unbroken four-arm bridge,

$$V_o = V_s G \varepsilon$$

which yields

$$0.15675 = (5)(2.090)\varepsilon$$
$$\varepsilon = 0.015$$

Now we can find the values of the two gauges on the left:

$$R_{gt} = R_g + G \varepsilon R_g, \quad R_{gt} = R_g - G \varepsilon R_g$$
$$R_{gt} = 123.76 \ \Omega, \quad R_{gb} = 116.24 \ \Omega$$

From these comes V_+, which is the voltage divider on the left with resistances R_{gt} and R_{gb}:

$$V_+ = 5 \frac{R_{gt}}{R_{gt} + R_{gb}} = 5 \frac{123.76}{116.24 + 123.76}$$
$$= 2.5783 \ \text{V}$$

Because the voltage on the right, V is 0, the voltmeter must be reading $V_o = 2.5783$ V.

3.4.2   EXAMPLE VI–RTD WIRING

An engineer you work with needs to know the wire resistance of the cable that ties an RTD to a control room display. The RTD is in a vat; its wires run in a standard instrumentation cable by some route that no one can locate. The only thought is that it's in a conduit somewhere.

To make this messier, the process that's running cannot be stopped. That means that your friend cannot simply apply an ohmmeter to the cable to see what its resistance is. An ohmmeter cannot be used when the circuit being checked is powered or is connected to something else.

Your friend devises a scheme where he is going to measure the temperature at the RTD tip using a simple thermocouple, then relate that temperature information to the voltage being read in the control room. He builds a Type E thermocouple, welding the twisted wires to make sure they stay together. Then he insulates them with a thin plastic coating. The lead wires are long enough to reach out of the vat so the second junction can be in ice water and he can attach a precision voltmeter to the thermocouple wires.

He positions one junction in the vat very close to the RTD tip. He positions the other in ice-water made with distilled water and carefully stirred to make sure the ice-water junction is truly at 0°C. He checks the ice water using a precision thermometer.

As he takes a reading of the thermocouple voltage, he asks the technician in the control room to take a reading of the voltage at that end of the RTD cable. Here are his data:

- Thermocouple voltage $V_T = 2.1256$ mV

- RTD voltage $V_R = 60.429$ mV

- RTD supply voltage (three-arm bridge) $V_s = 2.00$ V

The solution requires three steps. The first is to determine the temperature being sensed by the thermocouple, which he'll do using the NIST thermocouple polynomial for a Type E thermocouple shown in the table of Fig. 3.19. To avoid loss of precision, he's going to work to the nearest thousandth. The polynomial is

$$\Delta T = \sum_{i=0}^{n} a_n V_T^n$$

(V_T must be in microvolts.)

Using the numbers given in the table and the value of V_T from the engineer's measurement, the result is

$$\Delta T = 1.7057035 \times 10^{-2} V_T - 2.3301759 \times 10^{-7} V_T^2 +$$
$$6.5435585 \times 10^{-12} V_T^3 - 7.3562749 \times 10^{-17} V_T^4$$
$$= 36.256 - 1.053 + 0.063 - 0.002$$
$$= 35.264°C$$

The second step is to write the equation for the voltage from the 3-wire bridge using the temperature just measured and then equate this to the voltage V_R observed at the bridge:

$$2 \frac{(0.385)(35.264)}{400 + (0.77)(35.264) + 4R_w} = 60.429 \times 10^{-3}$$

Solving this for the wire resistance gives a value of $R_w = 5.547$ Ω per conductor.

Odd approach? Perhaps. Can you think of a different way that might not be as hard to carry out? Your goal is to obtain the resistance of the wires in the cable that connects the RTD through a

3-wire bridge to the control room without disrupting operations. [Clue: There is a very simple way that'll make you wonder where the engineer went to college!]

Another question. These results are supposed to be to the nearest thousandth. By how much is the final result changed if the ice-water junction is at 0.001°C instead of 0?

3.5   SUMMARY

On the surface, this chapter is about three particular instruments, the strain gauge, the resistance temperature detector, and the thermocouple. Knowing a little about how they work and how they are employed is only part of the story, though.

A larger and more important part is how the concept of measuring change around zero is done. The Wheatstone bridge is the key, because it effectively subtracts out of the equation the base value of a quantity, leaving us to observe only the changes. The bridge balances out the voltage that results from zero strain or from zero temperature. Then what we see are changes from that zero, sometimes called *difference measurement*.

FORMULAS AND EQUATIONS

1. Change in resistance—gauge factor G, strain ε, gauge resistance R

$$\Delta R = G\varepsilon R$$

2. Strain gauge two- and four-arm bridge

$$V_{o2} = \frac{V_s}{2}G\varepsilon, \quad V_{o4} = V_s G\varepsilon$$

3. Resistance temperature detector, DIN coefficient

$$R_g = 100(1 + 0.00385T) \quad \text{for T in } °C$$

4. Two- and three-wire RTD bridges

$$V_{2w} = V_s \frac{0.385T + 2R_w}{400 + 0.77T + 4R_w}$$

$$V_{3w} = V_s \frac{0.385T}{400 + 0.77T + 4R_w}$$

5. RTD four-wire ohmmeter

$$V_m = I_s R_g = (I_s)(100)(1 + 0.00385T)$$

6. Thermocouple polynomial, using table of Fig. 3.19

$$\Delta T = \sum_{i=0}^{n} a_i V_s^i$$

CHAPTER 4

Processing Signals

This stuff called electricity moves around for basically two purposes, moving energy or carrying information. The moving-energy part is in *Pragmatic Electrical Engineering: Fundamentals.* Here we'll look at carrying information.

We talk about carrying information in terms of *signals,* which are electrical transmissions that somehow have information encoded in them. There are lots of kinds of signals: telegraph, telephone, microwave, audio, cell phone, and many more.

This chapter is "old news" in a way, because many, many signals are digital in form. They are processed digitally, controlled digitally, filtered digitally, and so on. So what is this "old news?" We are going to restrict our signals visit to analog signals and analog transmission and analog filtering and analog processing. This means that we will use ordinary circuits as we have already learned them, circuits involving wires and voltage sources and capacitors and op-amps. Digital processing is beyond us here.

To put some of this into perspective, when you play music on a CD, your player is reading a digital bit stream from the CD. That stream has had encoded into it the digitally-sampled music, sampled 44,100 times per second. This bit stream is processed digitally—up to a point. For example, there is some buffering of the bit stream to "keep ahead" of where you are listening. This allows some error correction in case the CD slips.

At some point, though, that bit stream from the CD must be converted to analog form for the simple reason that our ears cannot make musical sense out of the bit stream. Part of that digital-to-analog conversion is filtering, because the analog signal must be limited to the band of frequencies between 0 and about 20 kHz. The analog signal also must be amplified to be able to drive your ear buds or speakers.

As I said, we'll stick to the analog part from here on. In the next chapter we'll take a look at how signals get interfered with and consider some things we can do to keep that interference to a minimum.

4.1 SIGNALS

How should we describe a signal? There are two fundamental ways. We can either state what the signal is doing in the time domain or in the frequency domain. Both are valid and both are useful, depending on what we are trying to do. In the time domain, we state what is going on as a function of time. In the frequency domain, we give the *spectrum* of the signal, information as a function of frequency. In this section we will do both.

4.1.1   SIGNALS IN THE TIME DOMAIN

A really simple signal is

$$v(t) = V_m \cos(2\pi ft + \phi)$$

But is this a signal? Does it convey information? Sure, of course, it obviously does. Doesn't it? Umm, how does it convey information? Oh, because it is either on or off and that is information, Morse code, perhaps. Pretty basic information, but information nevertheless.

Whoa! I didn't say it could be off or on! It's a sinusoid that is "on"—none of this on-or-off business! If this signal is on all the time, it conveys no information. As soon as you start turning it on and off, you increase its *bandwidth,* which means it is more than just a simple sinusoid at one frequency. For example, if you turn this sinusoid on and off in the manner of Morse code at 10 words per minute, the bandwidth required spreads out to something in the order of $f \pm 50$ hertz to be able to convey information.

Consider the square wave in Fig. 4.1. Its fundamental frequency is 1 kHz and its peak amplitude is 10 volts. I'll use this instead of Morse code to keep things simple. What sort of bandwidth does this square wave occupy?

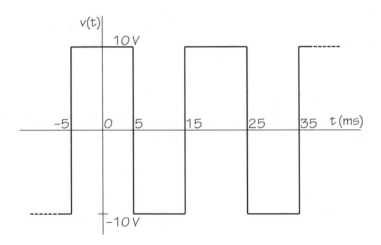

Figure 4.1: Square wave

Somewhere in your memory is the Fourier series. Remember that it says that any periodic signal can be represented as a series of sinusoids. The fundamental frequency within this series of sinusoids is the frequency of the periodic signal. Then there are *harmonics* of that fundamental frequency, integer multiples of the fundamental.

If I apply to this square wave what Fourier said, the result is

$$v(t) = 10 \sum_{\substack{k=1 \\ k \; odd}}^{\infty} \frac{4}{\pi k} \cos 2\pi k f t \quad \text{volts}$$

Yes, and...? What are we seeing here? The table below shows what is going on. The first column is number of the harmonic (the fundamental counts as $k = 1$), the second column is the actual frequency of the harmonic, and the last column is the peak value of that harmonic.

Harmonic k	Frequency kHz	Peak voltage
1	1.0	12.73
3	3.0	4.24
5	5.0	2.55
7	7.0	1.82
9	9.0	1.41
11	11.0	1.16
...	...	...
21	21.0	0.61

Yes, and ...? We started with a "simple" square wave at a frequency of 1 kHz and a peak amplitude of 10 volts. This square wave is made up of lots of frequencies, and they are not insignificant. For example, the 11th harmonic at 11 kHz has an amplitude of about 12% of the original square wave. At 21 kHz there's still a considerable voltage, over 6% of the original.

This time-domain representation of a signal looks like it gives us good information about the signal, but does it really? Sure, it gives us lots of numbers, but what do they mean? And what happens when the signal gets more complex, a voice, for example? This is where the *spectrum* comes in, or in other words, the *frequency-domain* description of the signal.

4.1.2   SIGNALS IN THE FREQUENCY DOMAIN

We often do not want to know the exact details of the sinusoidal content of a signal. We may not need the signal described as a function of time. It's enough to know the *spectrum* occupied by the signal. That information alone tells us a great deal about what we have to do to amplify, transmit, and receive the signal. We don't need its time-domain details.

We use spectrum information to describe audio equipment, for example. If you look up the specifications for a Klipsch KL-650-THX Bookshelf Speaker, made in Indianapolis at a price (MSRP) of $1,499, you find the statement "48 Hz – 20 kHz (±3 dB)." This means that if you draw a graph of the sound output of this speaker on a horizontal axis of frequency, you'll have a fairly flat straight line from 48 Hz to 20 kHz. The "fairly flat" can have an excursion of plus-or-minus 3 dB. (Some audio purists will tell you that they can hear, in that spectrum, some of those excursions.)

Radio Shack describes their 5-Element Slimline LCR Speaker ($629.99) as "80 Hz – 21 kHz." Any guess as to what that means? I have no idea how "flat" its curve is. (This is a plasma speaker; LCR means "left-center-right.")

The diagram of Fig. 4.2 is the spectrum of the response of the normal human ear. The vertical

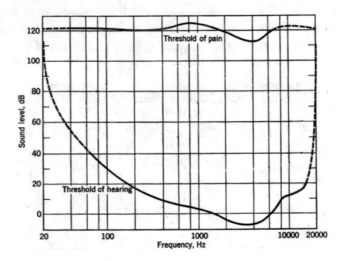

Figure 4.2: Sensitivity of human ear

unit is dB, but remember that dB requires a "zero" definition because it is always a ratio. Here, 0 dB corresponds to a sound pressure of 2×10^{-5} pascals (I'm sure you wanted to know!). The dotted portions of the curve are the regions generally outside of normal hearing. Notice in the figure that the ear's greatest sensitivity is around 3 kHz.

4.1.3   SIGNAL CONDITIONING

What might you want to do with a signal? The obvious choices are transmitting it and receiving it. But there are some other things we will want to do, though. One pretty clearly is to amplify it. The signal coming from the digital-to-analog converter in our CD player is very small but we want it to drive loudspeakers. We have to amplify that signal.

Another thing we often do to a signal is filter it. We want to change the spectrum of the signal, altering the distribution of energy as a function of frequency. For example, the telephone has a bandwidth of about 300 to 3,300 Hz for voice signals. If we are going to send an audio signal through the phone system, we will filter the signal to conform to the system's allowed bandwidth.

4.1.4   EXAMPLE I—MAKING A SIGNAL SMALLER

Suppose we want to control the *level* (amplitude) of a certain audio signal, a job we associate with a *volume control* or a *fader* in audio equipment. Figure 4.3 shows a device called a *potentiometer*, or

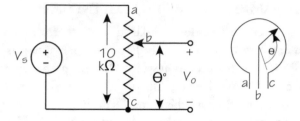

Figure 4.3: Potentiometer or "pot"

more commonly, a *pot*. It's set up as a voltage divider. The pot's resistance is spread uniformly over 330° of rotation. The slider rotates along that resistance to "pick off" a fraction of the resistance.

The resistance of that fraction of the total of 10 kΩ is

$$R_\theta = 10 \frac{\theta}{330°} \text{ k}$$

That makes the output voltage

$$V_o = \frac{R_\theta}{10} V_s = \frac{\theta}{330°} V_s$$

This simple device makes the signal amplitude controllable from nothing to full value. But voltage dividers must not leak, meaning there must be no current leaving the divider junction. So this simple pot-based volume control can't be connected to anything that requires current or the equation no longer applies.

Suppose this simple pot must be connected to an amplifier input that looks, to the pot, at least, like a 600-Ω resistor. This is a leaky voltage divider unless we isolate the divider from the amplifier's input. A voltage-follower op-amp circuit does this as in Fig. 4.4.

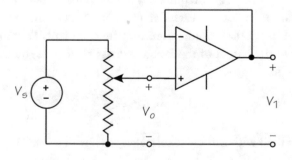

Figure 4.4: Pot with ouput isolated

4.1.5 EXAMPLE II—AMPLIFIER WITH LIMITATIONS

The maximum output of a certain transducer is 0.8 mV. In other words, when this transducer is stressed/pushed/heated or whatever it is doing, the largest output voltage is just 0.8 mV.

This transducer feeds an instrumentation system that, for proper operation, requires an input from 0 to ±0.5 V. This means we need an amplifier to take the small transducer signal and amplify it for the instrumentation system. Because this system is dealing with fast-changing conditions, we are told that the bandwidth of the amplifier should be at least 10 kHz. We think, OK, this sounds like a job for an op-amp.

Op-amp? Why not? The op-amp circuit must have a gain of

$$\frac{V_{inst}}{V_{transducer}} = \frac{0.5}{0.8 \times 10^{-3}} = 625$$

Gee, that's pretty large, but an inverting op-amp circuit could perhaps do this? Let's see. A gain of 625 would require 625 kΩ / 1 kΩ. That sounds suspicious because we have a general rule with op-amps that resistors should be in the range of "a k-ohm to a few hundred k-ohms." But 625 kΩ isn't too wild.

There is a worse problem hiding here, though—bandwidth. Amplifiers tend to have a response that "rolls off" at higher frequencies. In other words, the higher the frequency, the smaller the gain. This characteristic is often expressed as the *gain bandwidth product*. It's a number that says, the larger the gain, the smaller the bandwidth.

The gain-bandwidth product for a common op-amp such as the TL072 is 3×10^6 Hz. (Note that gain is unitless and bandwidth is frequency.) For this gain-bandwidth product, the bandwidth for a gain of 625 comes out

$$f_{bandwidth} = \frac{3 \times 10^6}{625} = 4.8 \text{ kHz}$$

That's way under 10 kHz that we require. So this amplifier arrangement doesn't do the job.

Let's put two op-amps in cascade, each one providing part of the required gain. The gains of amplifiers in cascade multiply if the amplifiers are truly independent. That means that connecting a second amplifier's input to the output of the first one doesn't affect the output of the first one.

The required gain per stage is

$$A_{stage} = \sqrt{625} = 25$$

This circuit can be built with two identical stages, each with a 1-kΩ input resistor and a 25-kΩ feedback resistor. The result is Fig. 4.5. The input of the second op-amp looks to the first one like 1 kΩ, which is not a significant load for an op-amp's output.

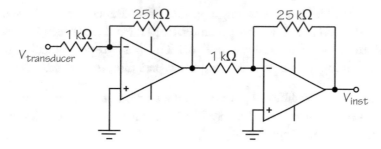

Figure 4.5: Cascade amplifiers

4.2   FILTERS

We've talked about signal conditioning already, or at least part of it. So far, we can make signals smaller and larger, and we can adjust their amplitudes. There's another major way that we condition signals—filtering. Signals often contain "parts" that we don't want. An example is a signal that contains, in the form of interference, a significant amount of 60-Hz noise.

What's the difference between noise and signal? Signal is the stuff we want and noise is the stuff we don't want. What's one person's signal is another person's noise. If you are listening to your favorite music in one room and your friend is listening to different music in another room, to each of you, your music is your signal. But if your friend's music is too loud and you are hearing it, that's noise to you.

How do you eliminate that noise? How do you eliminate your friend's music to keep it from interfering with yours? Close the door! But does that eliminate the noise? Or just reduce it to where it doesn't bother you? Probably the second. That's filtering.

We are going to keep our study of filters fairly simple. Our filters will all be *one-pole* filters. That means that all our filters denominators that have a single term of the form $(1 + s/a)$. Remembering that if we replace $s = j\omega$, this term becomes $(1 + j\omega/a)$. We'll see how to use this shortly.

4.2.1   BASIC FILTER FORMS

There are four basic forms of one-pole filters:

Low-pass (LP): This filter passes low frequencies and attenuates high frequencies. Example III (Chapter 1, Section 1.2.3) is a low-pass filter with a break frequency of 2,000 radians per second and a pass-band gain of 6 dB.

High-pass (HP): This filter passes high frequencies and attenuates low ones. Example IV (Chapter 1, Section 1.2.4) is a high-pass filter. Its break frequency is 500 radians per second; its pass-band is gain is 14 dB.

Band-pass (BP): This filter passes only a portion of the frequencies in the signal. Example V (Chapter 1, Section 1.2.5) is a band-pass filter. Its pass band is between 150 and 2,500 radians per second. In its pass band, the gain is 27 dB. This is a two-pole filter, but we'll design ours in this chapter using a high-pass filter and a low-pass filter in cascade.

Band-stop or *notch:* This filter stops or cuts out a portion of the frequency spectrum. A good notch filter is more complicated because it generally needs multiple poles. We are not going to deal with one in this chapter.

4.2.2   BREAK FREQUENCY AND HALF-POWER

The relationship between the break frequency, the pole, and the actual filter performance is an important one. Consider a filter whose voltage transfer function is

$$H(s) = \frac{V_{out}}{V_{in}} = \frac{k}{1 + \frac{s}{a}}$$

Replace s by $j\omega$:

$$H(s) = \frac{V_{out}}{V_{in}} = \frac{k}{1 + \frac{j\omega}{a}}$$

We evaluate this at $\omega = a$, which we should remember from Bode diagrams as the *break frequency*:

$$H(s) = \frac{V_{out}}{V_{in}} = \frac{k}{1 + \frac{ja}{a}} = \frac{k}{1 + j1}$$

The magnitude of this result is important:

$$|H(s = ja)| = \frac{k}{|1 + j1|} = \frac{k}{\sqrt{2}} = 0.707k$$

In dB, the magnitude of H at this break frequency is

$$|H(s = ja)|_{dB} = 20\log\frac{k}{\sqrt{2}} = 20\log k - 20\log\sqrt{2}$$
$$= k_{dB} - 3 \ dB$$

This says the actual magnitude of the filter's response is 3 dB inside the Bode asymptotes at the break frequency.

The magnitude at the break frequency is $k/\sqrt{2}$. Since the transfer function H(s) is a voltage ratio, this $\sqrt{2}$ says that the voltage is smaller by a factor $\sqrt{2}$. Because power is proportional to voltage squared, the power passing through the filter at this break frequency must be smaller by a factor $(\sqrt{2})^2 = 2$. In other words, the power is half as large as in the passband. This leads us to call the break frequency the *half-power point*.

In our original H(s), the frequency $s = j\omega = ja$ is important:

- $\omega = a$ is the break frequency where the Bode asymptotes meet.

- At $\omega = a$, the denominator of H(s) is $1 + j1$.

- If the denominator is $1 + j1$, the magnitude of the denominator is $\sqrt{2}$.

- This $\sqrt{2}$ is 3 dB, so the actual response curve is 3 dB inside the Bode asymptotes.

- $\sqrt{2}$ is in the denominator, so the power $(\sqrt{2})^2 = 2$ is half as much, which is why we call the break frequency the half-power point.

This will all hang together better with some examples.

4.2.3 EXAMPLE III—LOW-PASS PASSIVE FILTER

The circuit of Fig. 4.6 is a single-pole, passive, low-pass filter. We are going to be interested in the

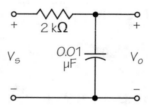

Figure 4.6: Low-pass passive filter

filter's transfer function H(s) = V_o/V_s and how those characteristics come about:

- There is going to be just one pole (a root of the denominator) in H(s) because there is just one energy-storage element, the capacitor. This will yield just one term in s

- A passive filter, or more generally, a passive circuit is one with no amplifier, no active element. It's constructed with just R, L, and C elements. As we'll see soon, though, we generally are going to want an amplifier somewhere in the circuit.

- How can we know that it is low-pass? Look at where the capacitor is in the circuit. Then recall that the impedance of a capacitor, 1/sC, decreases in magnitude as the frequency increases. At d–c, the capacitor looks like an open circuit and has no effect on the voltage. As the frequency increases, the capacitor's impedance decreases, which in effect begins to "short out" the output terminals. Hence, as frequency increases, output decreases.

It's probably easiest to find the transfer function H(s) and then the Bode diagram by using the voltage-divider relationship. The impedance of the capacitor is

$$Z_C = \frac{1}{0.01 \times 10^{-6} s}$$

Now write the voltage divider for H(s):

$$H(s) = \frac{V_o}{V_s} = \frac{Z_C}{R + Z_C}$$

$$= \frac{\dfrac{1}{0.01 \times 10^{-6} s}}{2000 + \dfrac{1}{0.01 \times 10^{-6} s}} = \frac{1}{1 + 20 \times 10^{-6} s}$$

This needs to be in the standard form for making Bode diagrams as we did in Section 1.2. All terms involving s need to be written as $1 + s/a$:

$$H(s) = \frac{V_o}{V_s} = \frac{1}{1 + \dfrac{s}{50 \times 10^3}}$$

The Bode diagram has two asymptotes:

- The numerator is 1, which in dB is $20 \log 1 = 0$ dB, so this asymptote is a horizontal line at 0 dB.

- The denominator has a break frequency of 50×10^3 radians/second. To the left of that break, the curve is a horizontal line at 0 dB; to the right, the curve slopes downward at 20 dB/decade.

These asymptotes are logarithms (dB), so we combine them by addition. The result is the Bode diagram of Fig. 4.7. It's flat at 0 dB up to $\omega = 50$ krad/s. Then it slopes downward at 20 dB per decade. This diagram also includes the smooth line of the actual curve as well as the asymptotes. Notice that the smooth curve is 3 dB "inside" the asymptotes at the break frequency. (It's inside by 1 dB at half and twice the break frequency.) This *half-power point* or *3 dB point* comes from

$$\left| H\left(s = j50 \times 10^3\right) \right| = \left| \frac{1}{1 + \dfrac{j50 \times 10^3}{50 \times 10^3}} \right| = \left| \frac{1}{1 + j1} \right| = \frac{1}{\sqrt{2}}$$

$$\left| H\left(j50 \times 10^3\right) \right|_{dB} = 20 \log \frac{1}{\sqrt{2}} = -3 \ \text{dB}$$

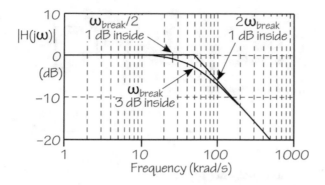

Figure 4.7: Bode plot for Example III

There! A passive low-pass filter! Now what are we going to do with it? I presume we had a reason for filtering the particular signal V_s, so we must have something in mind that we are planning to pass the signal to. But what?

Umm, OK, how about feeding another circuit, maybe a controller? Sure, but let's be careful. We analyzed the filter by using the voltage-divider relationship, which requires that the divider not "leak." There can be no current flowing to the right from the R-C node. So what circuit can we connect to?

How about the voltage follower in Fig. 4.8? The op-amp's input circuit draws essentially no

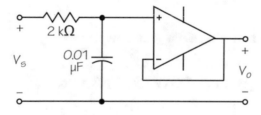

Figure 4.8: "Improved" low-pass filter

current, so it won't mess up the voltage divider. Philosophically, this seems to violate the *passive* idea.

Well, while we are at it, how about putting in an op-amp circuit with gain? This way, we not only filter the signal but we amplify it. OK, sounds good. Let's add a gain of 2 to see what happens. I'll add an inverter with a 1-kΩ input resistor and a 2-kΩ feedback resistor as shown in Fig. 4.9.

NO! NO! That won't do! The input to the inverter draws current through the 1-kΩ resistor. This has altered the voltage divider. The inverter won't work.

The non-inverter will do the job as shown in Fig. 4.10. Its input terminal draws almost no current, so it doesn't affect the divider. So now we have an *active* low-pass filter with a pass-band gain of 2 and a cutoff frequency of 50×10^3 radians per second.

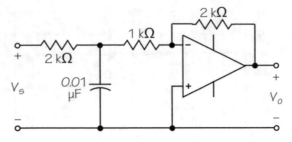

Figure 4.9: INCORRECT low-pass filter

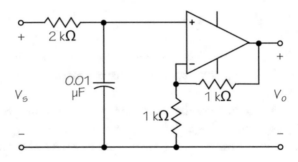

Figure 4.10: Better low-pass filter

There's a better way to do this, as the next example shows.

4.2.4 EXAMPLE IV—ACTIVE LOW-PASS FILTER

This time I want an active low-pass filter. Its break frequency is to be 50×10^3 radians/second. Its pass-band gain is to be 2. This can be done using the inverter of Fig. 4.11. Here's how it works:

- The R and the C in the feedback path provide the filter aspect of the circuit. These are the same values as in the previous example and provide the same break frequency.

- At very low frequencies such as d-c, the capacitor is not a factor, so this inverter has a low-frequency gain of -2 via the two resistor values.

Analysis of this circuit is most easily done by first finding the impedance of the feedback path and then using our knowledge of the inverter to get H(s).

The impedance of the feedback path is

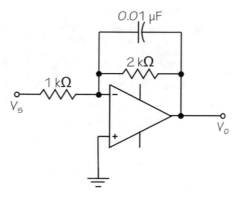

Figure 4.11: Even better active low-pass filter

$$Z_f = \frac{1}{0.01 \times 10^{-6}s} \| 2000 = \frac{1}{0.01 \times 10^{-6}s + \dfrac{1}{2000}}$$

$$= \frac{2000}{1 + \dfrac{s}{50 \times 10^3}}$$

Using what we know about the inverter:

$$H(s) = \frac{V_o}{V_s} = -\frac{Z_f}{1000} = \frac{-2}{1 + \dfrac{s}{50 \times 10^3}}$$

The Bode diagram (Fig. 4.12) has an asymptote for the "2" at 6 dB and a break downward at 50×10^3 radians/second.

4.2.5   RC AND BREAK FREQUENCY

Notice how the product RC keeps appearing in the equations for these filters. In general, an R that can be "seen" by a C sets the break frequency. That frequency will be

$$\omega_{break} = \frac{1}{RC}$$

$$f_{break} = \frac{\omega_{break}}{2\pi} = \frac{1}{2\pi RC}$$

What does "seen" mean? If an R and a C are in parallel, C sees that R. If they are in series, likewise, C sees that R. So we can generally determine the break frequency of one-pole filters by looking at the capacitor and noting what resistance it sees.

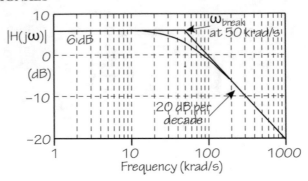

Figure 4.12: Bode plot for Example IV

4.3   MORE EXAMPLE

In the examples that follow, we'll design active filters to meet frequency and gain specifications. I'm not going to attempt to use standard values of resistors and capacitors. If these filters are to be built, the element values will probably need to be altered to fit what is available. Resistor values present little problem because the 1%-series of resistances can handle all but the most demanding precision for a filter. Capacitors are harder to find with precise values.

4.3.1   EXAMPLE V—LOW-PASS ACTIVE FILTER

Design a low-pass filter with the following specifications:

- Cutoff (break) frequency of 4 kHz

- Pass-band (low-frequency) gain of 8 dB

- Input resistance at least 1 kΩ

I'll start by choosing the input resistor to be 1 kΩ. This is simple and meets the third specification. That choice leads to the choice of the feedback resistor, which will determine the low-frequency gain.

The low-frequency or pass-band gain is given in dB, so it needs to be changed to a plain number A:

$$8 \ dB = 20 \log A$$

Turning this around to find A:

$$\frac{8}{20} = \log A$$
$$10^{8/20} = 10^{\log A} = A$$
$$A = 10^{8/20} = 2.5$$

So if the input resistor is 1 kΩ, the feedback resistor needs to be 2.5 kΩ to provide a gain of 2.5. (Actually, the gain is −2, but we'll ignore the inversion.)

The break frequency or the cutoff frequency (two names for the same thing) is determined by 1/RC in radians/second, where R and C are in parallel in the feedback path. R is already set to be 2.5 kΩ, so we need to find C, not forgetting to translate via 2π from hertz to radians/second:

$$\frac{1}{R_f C_f} = (2\pi)(4 \times 10^3), \quad R_f = 2.5 \times 10^3 \Omega$$

$$C_f = \frac{1}{(2.5 \times 10^3)(2\pi)(4 \times 10^3)} = 1.59 \times 10^{-8}$$

$$\approx 0.016 \ \mu F$$

This filter is in Fig. 4.13 and its Bode diagram is Fig. 4.14.

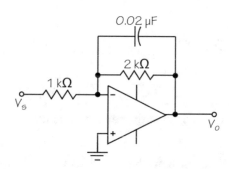

Figure 4.13: Low-pass filter—Example V

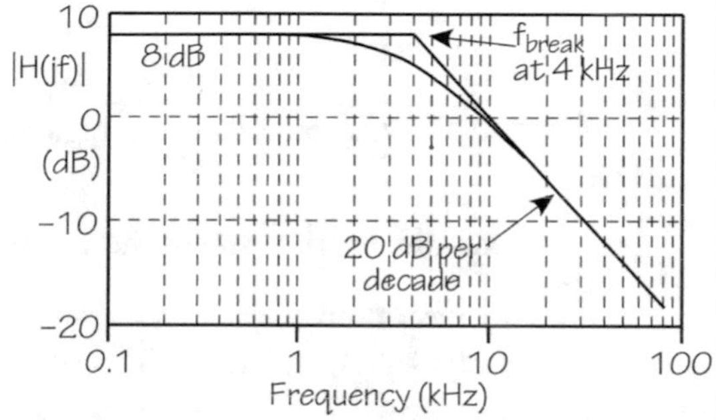

Figure 4.14: Bode plot for Example V

4.3.2   EXAMPLE VI—HIGH-PASS ACTIVE FILTER

Design a high-pass filter as follows:

- Cutoff frequency of 800 Hz

- Pass-band (high-frequency) gain of 1

This time the input resistance is not specified, so I'll just pick one—how about 1 kΩ? That means…whoa! Wait a minute. What is the circuit for a high-pass filter? The filter with a parallel RC in the feedback path is *low*, not high.

Suppose I put a capacitor *in series* in the input path as shown in Fig. 4.15. Now find H(s) for this circuit. The input impedance is

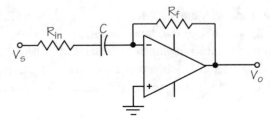

Figure 4.15: High-pass filter model

$$Z_{in} = R_{in} + Z_C = R_{in} + \frac{1}{Cs} = \frac{1 + R_{in}Cs}{Cs}$$

This filter's transfer function is

$$H(s) = -\frac{R_f}{Z_{in}} = -\frac{R_f}{\frac{1 + R_{in}Cs}{Cs}} = -\frac{R_f Cs}{1 + R_{in}Cs}$$

Look at the outcome in three regions, remembering that s is $j\omega$ when we do frequency calculations:

- The denominator determines the cutoff or break frequency, so $\omega_{cutoff} = 1/R_{in}C$

- For very low frequencies (s very small, a value much smaller than the cutoff frequency), $|H(s)|$ will be very small, and at d-c ($s = 0$), $H(s) = 0$

- For very high frequencies (s very large, a value much larger than the cutoff frequency), $|H(s)|$ becomes

$$|H(s)| \approx \frac{R_f Cs}{R_{in}Cs} = \frac{R_f}{R_{in}}$$

Now I know how to proceed. I've chosen the input resistance to be 1 kΩ, so to have a high-frequency gain of 1, I need to choose the feedback resistor to be 1 kΩ also.

The capacitor's value comes from

$$\omega_{cutoff} = (2\pi)(800) = \frac{1}{R_{in}C} = \frac{1}{1000C}$$

$$C = \frac{1}{(1000)(2\pi)(800)} = 1.99 \times 10^{-7}$$

$$\approx 0.2 \ \mu F$$

Figure 4.16 is the resulting circuit and 4.17 is the Bode diagram showing the asymptotes and the actual response curve.

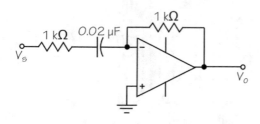

Figure 4.16: High-pass filter—Example VI

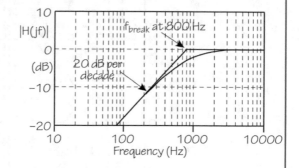

Figure 4.17: Bode plot for Example VI

4.3.3 EXAMPLE VII—BAND-PASS ACTIVE FILTER

Design a band-pass filter for voice in telephone circuits:

- Lower cutoff at 300 Hz

- Upper cutoff at 3.3 kHz

- Pass-band gain of 10 dB

I think a good way to start this one is to look at the Bode diagram that meets these specifications. See Fig. 4.18 and notice the following:

- There are two break frequencies, 300 and 3300 Hz

- Below 300 Hz, the slope is 20 dB/decade

- Above 3300 Hz, the slope is −20 dB/decade

- In the middle between the cutoff frequencies, the gain is 10 dB

A simple way to do this job is to consider this filter to be the combination of two that we already know about, a low-pass filter and a high-pass filter. Perhaps surprisingly, we can do this with just one op-amp as Fig. 4.19 shows. The series RC on the input provides the high-pass characteristic, while the parallel RC in the feedback path provides low-pass.

We need to be careful not to think backward, though! The high-pass characteristic provides the *300-Hz* break, while the low-pass characteristic provides the 3300-Hz one. It's easy to mix high and low here. These will yield two equations:

$$\omega_{lower} = \frac{1}{R_{in} C_{in}}$$

$$\omega_{upper} = \frac{1}{R_f C_f}$$

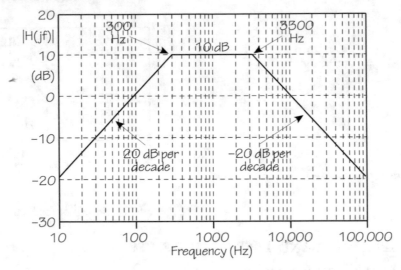

Figure 4.18: Specifications for Example VII

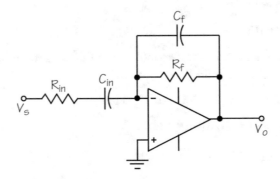

Figure 4.19: Band-pass filter model

(Be careful of radians and hertz!)

But this is just two equations and there are four elements. Now what? Well, consider the pass band between the cutoffs. That part of the curve is the pass band of both the high-pass and the low-pass filters. If we think back to those simple filters, in the pass bands their resistors dominated. In the pass band of both high-pass and low-pass filters, the input and feedback resistors set the gain.

The same is true here as long as the cutoff frequencies aren't too close together. A decade is about enough for "not too close." In this region, the gain of the band-pass filter is approximately $-R_f/R_{in}$. That gives me another equation, where the "10" in the exponent is my desired pass-band gain of 10 dB:

$$\frac{R_f}{R_{in}} = 10^{10/20} = 3.16$$

That's now three equations and four unknowns. What about the fourth equation? We don't have one, so we simply have to choose a value for one of the elements and go from there.

But which one? It turns out that capacitors are often the best elements to "just pick." I will pick the capacitor that is likely to be the largest one in the circuit. If I look at the two cutoff equations, I note that the lower frequency, the larger the capacitor. So I am going to choose

$$C_{in} = 0.1 \ \mu F$$

This value for C is about as large as practical for small-sized commercial capacitors.
That gives me the value for R_{in}:

$$R_{in} = \frac{1}{\omega_{lower}C_{in}} = \frac{1}{(2\pi)(300)(0.1 \times 10^{-6})} = 5.31 \ k\Omega$$

I can now calculate R_f:

$$\frac{R_f}{R_{in}} = 3.16$$

$$R_f = (3.16)(5.31 \times 10^3) = 16.8 \ k\Omega$$

Finally, I get C_f:

$$C_f = \frac{1}{\omega_{upper}R_f} = \frac{1}{(2\pi)(3300)(16.8 \times 10^3)}$$

$$= 2.88 \times 10^{-9} = 0.00288 \ \mu F$$

That does it and the result is the circuit of Fig. 4.20.

But does this circuit do the job? I'm going to finish all this by calculating the frequency response of this circuit using the actual element values and checking at least three points, namely, the values of the transfer function at the two break frequencies and at the middle of the pass band.

This calculation is most easily done by finding the impedances of the input and feedback paths first:

$$Z_f = \frac{1}{2.88 \times 10^{-9} s} \| 16.8 \times 10^3 = \frac{16.8 \times 10^3}{1 + 4.84 \times 10^{-5} s}$$

$$Z_{in} = 5.31 \times 10^3 + \frac{1}{0.1 \times 10^{-6} s} = \frac{1 + 5.31 \times 10^{-4} s}{0.1 \times 10^{-6} s}$$

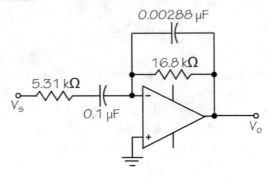

Figure 4.20: Band-pass filter—Example VII

Their ratio yields the overall transfer function:

$$H(s) = -\frac{Z_f}{Z_{in}} = -\frac{\dfrac{16.8 \times 10^3}{1 + 4.84 \times 10^{-5} s}}{\dfrac{1 + 5.31 \times 10^{-4} s}{0.1 \times 10^{-6} s}}$$

$$= -\frac{1.68 \times 10^{-3} s}{\left(1 + 4.84 \times 10^{-5} s\right)\left(1 + 5.31 \times 10^{-4} s\right)}$$

From the poles in the denominator, the two break frequencies turn out to be

$$f_{lower} = \frac{\omega_{lower}}{2\pi} = \frac{1}{2\pi}\left(5.31 \times 10^{-4}\right)^{-1} = 299.7 \ \text{Hz}$$

$$f_{upper} = \frac{\omega_{upper}}{2\pi} = \frac{1}{2\pi}\left(4.84 \times 10^{-5}\right)^{-1} = 3.288 \ \text{kHz}$$

How about the gain at the middle of the pass band? I'll chose the middle by eye to be about 1 kHz. The magnitude of the transfer function at this frequency is

$$|H(j\omega)| = \left|\frac{1.68 \times 10^{-3} j\omega}{\left(1 + 4.84 \times 10^{-5} j\omega\right)\left(1 + 5.31 \times 10^{-4} j\omega\right)}\right|$$

At $\omega = (2\pi)(1000)$, $|H(j\omega)| = 2.900 = 9.25 \ \text{dB}$

Figure 4.21 is the Bode diagram with the actual curve drawn on the desired asymptotes. You can judge whether this design does the job!

4.4 SUMMARY

We've processed a few signals, mostly by filtering, all by analog devices. We've done this filtering by using just two very straightforward one-pole filters, low-pass and high-pass. Each involves just

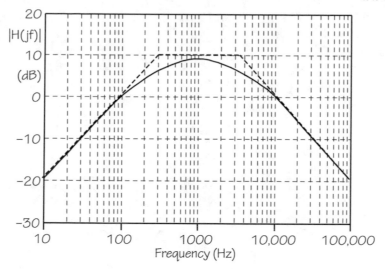

Figure 4.21: Actual plot for Example VII

one capacitor and one resistor, but each is made more versatile by adding one op-amp to make it an active filter. More extensive filters can be built up from these simple ones.

FORMULAS AND EQUATIONS

1. Break frequency for single R-C pair (either series or parallel)

$$\omega_{break} = \frac{1}{RC}$$

$$f_{break} = \frac{\omega_{break}}{2\pi} = \frac{1}{2\pi RC}$$

2. Low-pass active filter

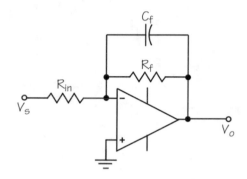

3. High-pass active filter

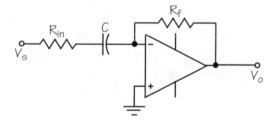

4. Band-pass active filter

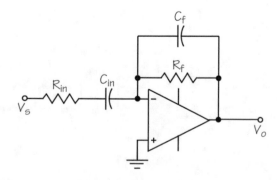

CHAPTER 5

Electromagnetic Compatibility

Huh? What's this electro…something-or-other? Well, how about noise NOISE NOISE? Calling this *electromagnetic compatibility* is just a way of stating the positive side of the problem rather than the negative. Instead of screaming at the kids to stop hammering on the coffee table, you arrange for something else they can bang on.

We live in a noisy world, not just audible noise but electrical noise as well. This electrical noise can interfere with the proper operation of electrical apparatus such as instruments, control systems, communications links, and computers. *Compatibility* implies that we design our systems so that electrical noise does not disturb them unduly—and our systems don't disturb others unduly, either.

Have you ever read the required statement on RF Emissions that comes with just about every piece of electronics today? It says in part, "Operation is subject to the following two conditions: (1) This device may not cause harmful interference, and (2) This device must accept any interference received, including interference that may cause undesired operation."

This says that we have to consider how to protect our systems from electrical interference so that they continue to operate properly in the face of that interference. In other words, let the kids keep banging on the coffee table, ignore the din, and go about your work.

In this chapter we look at the kinds of interference that our systems must put up with, how that interference is coupled into our systems, and how we might keep it out or at least diminish its effects.

5.1 NOISE

Noise is pretty easy to define. It's the same as *signal* but it's signal we *don't* want. What's one person's signal is another person's noise. Electromagnetic noise is unwanted energy that is conveyed electrically or magnetically into your system. It is a problem if it interferes with the proper operation of that system. Our job is to design our systems so that noise doesn't bother their functioning.

Figure 5.1 looks at a system as being a source of information, a channel over which the information is transmitted, and a destination that uses that information. This doesn't have to be a communications link like a telephone wire or a radio signal. It can be a memory chip that is communicating with a processor on the same circuit board.

Electrical noise can enter this system everywhere as the diagram shows. Noise is always present, so it's our job either to keep it out or to reduce its effect on our system. But we cannot keep it out 100%. Hence, we need to learn about how this noise gets in before we can figure out how to stop it.

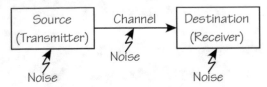

Figure 5.1: Interference

There are two general categories of electrical noise, and we deal with them in somewhat different ways:

Pulse noise: This noise, generally called an *electromagnetic pulse (EMP)* is sometimes mentioned in the media as the outcome of a nuclear explosion. But we see EMP quite often in the form of lightning. A lightning strike nearby produces a large pulse of energy that can couple into our systems and destroy components. This can happen even if the strike is merely nearby, not direct.

There are other common sources of pulse noise. Direct-current motors are a prime source, where the brushes switching from one commutator segment to another cause noise. Power-line switching transients and switching power supplies (in every computer) are also sources.

Continuous noise: An obvious source is a nearby radio transmitter, some of whose energy couples into our system. Any piece of 60-Hz equipment is a source of possible 60-Hz noise, even the lights in the room.

Noise gets into our systems through three mechanisms:

- *Capacitive coupling* is simply a capacitance that is formed between two conductors. Since a capacitor is just two metal plates separated by a dielectric, two wires near one another can be a capacitor. There's nothing that says the metals have to be flat plates.

- *Magnetic coupling* comes about because a current in a wire or coil of wire creates a magnetic field that can couple into another conductor.

- *Electromagnetic radiation* enters "over the air" in the same way broadcast signals come from radio stations.

Our job is to understand a bit about these three mechanisms. Our goal, of course, is to keep the energy out, but that's not 100% possible, so we will seek ways to diminish the unwanted effects.

5.2   CAPACITIVE COUPLING

Since capacitive coupling implies the existence of capacitance, we can model this form of interference by a simple circuit with a capacitor "connection." There are two circuits in Fig. 5.2. The one on the left is the source of the noise and is modeled as just a voltage source. The one on the right, often called the *victim circuit*, is receiving the noise. The capacitor over the top represents capacitive coupling.

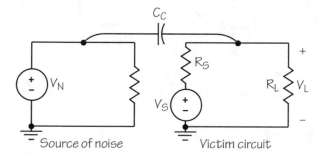

Figure 5.2: Capacitive coupling

5.2.1 CALCULATING CAPACITIVE COUPLING

In order to understand the effects of this coupling, we need to see how much of the noise voltage makes its way into the victim circuit. If we assume that both of these circuits are linear, we can apply the principle of superposition. This principle says that we can look at the effects of just one source at a time while insuring that the other sources are dead.

If we apply superposition to the circuit of Fig. 5.2, we leave the noise-voltage source V_N turned on and we turn off the signal source V_S. "Turning off" means we replace the source with a short circuit to guarantee that the voltage where it was is zero. The result is the circuit of Fig. 5.3.

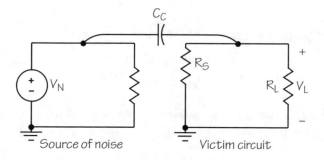

Figure 5.3: Victim with $V_S = 0$

The two resistors of the victim circuit are in parallel, so I'll reduce the circuit to the only elements affected by the noise voltage (Fig. 5.4): the source of the noise V_N, the coupling capacitor C_C, and the resistors of the victim circuit, labeled R_O. (R_O is R_S in parallel with R_L.)

The circuit is a voltage divider, so I can figure out the voltage V_L coupled into the victim:

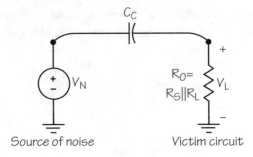

Figure 5.4: Noise coupling only

$$V_L = V_N \frac{R_O}{R_O + \dfrac{1}{j\omega C_C}} \quad \text{where } R_O = R_S \| R_L$$

This twists around to become

$$V_L = V_N \frac{1}{1 + \dfrac{1}{j\omega C_C R_O}}$$

The magnitude of the coupling is what we want to look at:

$$\left| \frac{V_L}{V_N} \right| = \frac{1}{\sqrt{1 + \dfrac{1}{\omega^2 C_C^2 R_O^2}}}$$

Consider what happens at the ends of the frequency range:

$$\left| \frac{V_L}{V_N} \right| = \begin{Bmatrix} 0 & \omega = 0 \\ 1 & \omega \to \infty \end{Bmatrix}$$

OK, are you ready for an obvious statement? Coupling gets worse as the frequency goes up. So if that's the case, what do we do to reduce this interference? It's not hard to figure out what to attack:

- The operating frequency of our circuit is fixed by the requirements of whatever we are doing. So the frequency ω is fixed.

- The impedance of our victim circuit (here, R_O) is fixed by the needs to our circuit. We perhaps could lower the impedance because making it smaller makes the denominator of $|V_L / V_N|$ larger and reduces the coupling. But we might not have that choice because of the design of our circuit.

- That leaves just the coupling capacitance C_C to adjust. Reducing the coupling capacitance makes the denominator of $|V_L / V_N|$ larger and reduces the voltage coupled into our victim circuit.

Sure, this outcome should have been obvious! If you want to reduce the capacitive coupling, reduce the coupling capacitance! By going through the calculations, though, we did see what options might have been available to us.

5.2.2   REDUCING CAPACITIVE COUPLING

Rather than looking at the capacitance between two wires, which is a messy equation, let's look at the simpler case of a parallel-plate capacitor:

$$C = \frac{\varepsilon A}{d}$$

This says that capacitance is proportional to the area of the plates. Since the "plates" in our interference problem are really wires that have been chosen for being the proper ones for whatever we are doing, we can't change those.

The equation also says that capacitance is inversely proportional to the distance between the plates. That's our clue as to how to reduce capacitive coupling—separate the conductors further. (OK, if you were really expecting something more complicated, sorry!)

But surprisingly, there is another way: add capacitance to the circuit. Yes, add! But add this capacitance in a special way. We do this by adding a conducting ground plane near the wires that are interfering with one another.

The drawing of Fig. 5.5 shows two wires end on. Between the two wires is the coupling

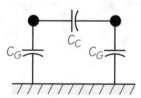

Figure 5.5: Conductors over ground

capacitance C_C. From each of the wires to the ground plane is the ground capacitance C_G. Even though this seems strange, it works. In an overly simple way, the ground plane "diverts" some of the electric field away from coupling the two conductors.

I'll add the noise voltage source V_N and the victim's resistance R_O to the drawing (see Fig. 5.6). Then I can analyze the circuit as before. The voltage divider consists of R_O in parallel with C_G as one leg and the coupling capacitance C_C as the other:

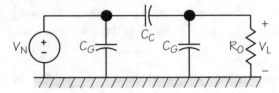

Figure 5.6: Conductors over ground in circuit

$$V_L = V_N \frac{1}{1 + \dfrac{1}{j\omega C_c Z_o}} \quad where \quad Z_o = R_o \left\| \frac{1}{j\omega C_G} \right.$$

The noise coupling will be reduced, meaning that will be smaller if

$$|j\omega C_c Z_o| < |j\omega C_c R_o|$$
$$|Z_o| < |R_o|$$

Adding C_G in parallel with R_O does yield an impedance whose magnitude is smaller than R_O, so it looks like this will work. Let's try some numbers.

5.2.3   EXAMPLE I–BENEFIT OF GROUND PLANE

Two conductors are spaced 3 mm apart (Fig. 5.7). There is no ground plane nearby. The circuits

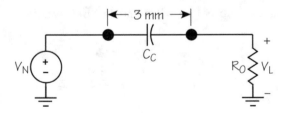

Figure 5.7: Parallel conductors coupling

have been simplified to just the noise source, the coupling capacitor, and the resistances of the victim circuit.

The wires are #22 American Wire Gauge (AWG), whose diameter is 0.644 mm. Both wires are one meter long. Capacitance per meter of two wires in parallel is

$$C_c = \frac{27.78}{\ln(d/r)} \text{ pF/m}$$

where d is the spacing of the wires and r is the radius of the wire. Using this formula, the coupling capacitance is

$$C_C = \frac{27.78}{\ln(3/0.322)} = 12.4 \text{ pF/m}$$

The noise source is operating at 100 MHz (in the middle of the FM broadcast band). Both resistances R_S and R_L are 75 ohms, so $R_O = 37.5\Omega$. From the voltage divider that we developed in Section 5.2.1, the noise voltage coupled into the victim circuit is

$$\frac{V_L}{V_N} = \frac{1}{1+\dfrac{1}{j\omega C_c R_o}} = \frac{1}{1+\dfrac{1}{j(2\pi)(10^8)(12.4\times 10^{-12})(37.5)}}$$

$$= 0.280\angle 73.7°$$

Now let's see what adding the ground plane does to this. The circuit of Fig. 5.8 shows the circuit with the capacitances to the ground plane. The conductors are each 0.2 mm from the ground plane.

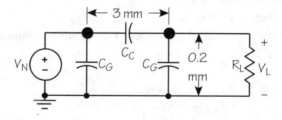

Figure 5.8: Conductors with ground plane

The formula for capacitance per meter of a wire over a ground plane is

$$C_c = \frac{27.78}{\ln(2h/r)} \text{ pF/m}$$

where h is the height of the wire above the plane and r is the radius of the wire. This formula gives a ground capacitance of

$$C_c = \frac{27.78}{\ln(2(0.2)/0.322)} = 128.1 \text{ pF/m}$$

In the circuit of Fig. 5.8, the capacitor on the left has no effect because the voltage source is directly across it. The voltage divider is a resistor and capacitor for one leg and the coupling capacitor for the other. The impedance of the vertical leg is

$$Z_O = \cfrac{1}{\cfrac{1}{37.5} + j(2\pi)(10^8)(128.1 \times 10^{-12})}$$

$$= 11.8 \angle - 71.7°$$

The noise voltage coupled into the victim circuit is now

$$\frac{V_L}{V_N} = \cfrac{1}{1 + \cfrac{1}{j(2\pi)(10^8)(12.4 \times 10^{-12})(11.8 \angle -71.7°)}}$$

$$= 0.0845 \angle 16.8°$$

Did this do any good? The added capacitance to the ground plane drops the coupling from 0.280 to 0.0845. That's 10.4 dB. That's one reason why circuit boards for high-speed logic (like computers) usually include a ground plane between layers of circuit traces.

5.3   MAGNETIC COUPLING

Magnetic coupling is another mode by which energy sneaks from one circuit to another, bringing with it unwanted energy that we call noise.

Figure 5.9 shows a conductor carrying a current, along with the resultant magnetic flux density B. The magnitude of this flux density is

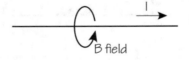

Figure 5.9: Field around conductor

$$|B| = \mu|H| = \frac{\mu I}{2\pi r}$$

Coupling a voltage into another circuit depends on the total flux:

$$V_N = \frac{d\psi}{dt} = \frac{d(BA)}{dt}$$

If the area A is constant, this becomes

$$V_N = A\frac{dB}{dt}$$

Since the only thing changing is the current, the coupled noise voltage is

$$V_N = \frac{\mu A}{2\pi r}\frac{dI}{dt}$$

Now suppose the current is sinusoidal:

$$I = I_o \cos\omega t$$

Then the noise voltage is

$$V_N = \left(-I_o\omega\sin\omega t\right)\left(\frac{\mu A}{2\pi r}\right)$$

whose peak magnitude is

$$|V_N| = \frac{\mu A}{2\pi r}I_o\omega = |B|A\omega$$

How do we reduce this voltage? We could change the frequency ω but that's generally fixed by the signal requirements, so we are left with three other choices:

- Reduce the area of the victim circuit that is being presented to the magnetic flux

- Increase the distance between the noise source and the victim circuit, thereby reducing B

- Change the angular relationship between the two circuits, since magnetic coupling depends very much on this relationship (although we have left it out of the analysis)

The third method is one that isn't hard to do. Consider the cable of Fig. 5.10. The output

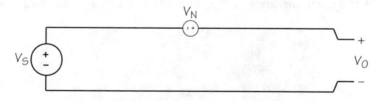

Figure 5.10: Parallel conductors: 1 loop

voltage V_O will be the sum of the two voltage sources, signal and noise:

$$V_O = V_S + V_N$$

Now twist the wires so there are three loops (Fig. 5.11):
Two of the loops cancel each other, so let's keep going—how about six twists (Fig. 5.12)?

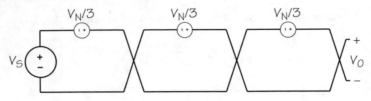

Figure 5.11: Parallel conductors: 3 loops

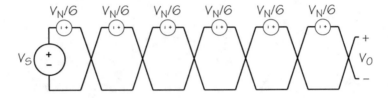

Figure 5.12: Parallel conductors: 6 loops

$$V_O = V_S + V_N/3 - V_N/3 + V_N/3$$

This sounds too good to be true, and it probably is. The common practice is to use a reduction by the number of twists:

$$V_O = V_S + V_N/6$$

Convention is that *n* should be limited to 1,000 twists; going beyond that yields unrealistic numbers.

Figure 5.13 is a communications cable. This one is a Belden cable with two twisted pairs

Figure 5.13: Communications cable

separated in foil shields. The conductors are #22 AWG. They are color-coded so that they can be easily identified. Notice especially the aluminized Mylar shields that surround the pairs. These provide ground planes and the separation between the pairs. The thin bare wire is a *drain* that provides a solid ground conductor through the cable.

Figure 5.14 is an Ethernet cable. This one is a Cat 5 cable. It has no shields, but it makes

Figure 5.14: CAT 5 cable

clever use of twists. Each of the four pairs is twisted, but the twists are different. The numbers of twists per inch are not integral multiples of one another so the twists of one don't stay in step with the twists of the others.

5.4   ELECTROMAGNETIC SHIELDING

The third mode by which energy enters our circuits is electromagnetic radiation. While that sounds dangerous, this is the non-ionizing radiation that is produced intentionally by every radio transmitter and unintentionally by just about every other electronic device. In the right places, this energy is beneficial, bringing information and music or cooking our food in the microwave oven. In our circuits, though, it is just noise.

We've considered how one circuit can affect another by both capacitive and inductive coupling. We've seen how one might separate circuits by adding ground planes or by changing orientation. Now we need to look at how we can keep this radiated energy out of our circuits. The answer is shielding.

A good example of the need for shielding is your computer. All computers generate a significant amount of electrical noise because of all the high-speed pulses in their circuits. Much of the shielding of computers is to keep their noise inside, not letting it out to affect other circuits and electronics. The Federal Communications Commission has very strict rules on how much electrical noise can be allowed to escape from electronic devices such as computers.

What does shielding mean? It means constructing a barrier between the noise source and the rest of the world. Figure 5.15 shows two circuits separated by a shield. Circuit 1 is radiating electrical energy. The shield is to keep that energy out of Circuit 2. Keep in mind that this energy is arriving not from capacitive effects and not from magnetic effects but directly by electromagnetic radiation.

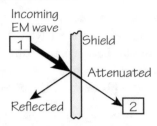

Figure 5.15: Shield effects

The shield keeps radiated energy out in two ways:

- *Absorption* reduces the energy as it passes through the shield. Some gets through, but reduced to what should be an acceptable level. I'll use the letter A in connection with this mode of reduction.

- *Reflection* bounces the energy back away from the shield, much like a mirror bounces light. I'll use R for this mode.

We'll measure shield effectiveness in decibels using voltage:

$$S = 20 \log_{10} \frac{V_{no\ shield}}{V_{shield}}$$

Writing this in terms of the two shielding mechanisms I've just stated yields the basic formula for effectiveness of a shield:

$$S = A + R \ \text{ in dB}$$

I'm dividing our look at shielding into three parts: absorption, reflection, and effects of holes in the shield. Holes are included separately because holes actually attenuation electromagnetic waves, too. As evidence of this, note the screen in the window of your microwave oven. Energy doesn't come through those holes.

As we go along, pay attention to notation and to the various constants involved. To help a bit, I've gathered some of the important constants into the table of Fig. 5.16 so that we have all of them in one place with their units.

5.4.1 ABSORPTION BY THE SHIELD

Absorption is not hard to visualize. As the electromagnetic wave strikes the surface of the shield, it begins to penetrate. As it gets deeper, it gets weaker as it gives up its energy to the material.

In the formula below, E is the strength of the electromagnetic wave represented in terms of voltage. Its unit is actually volts per meter, but this E will drop out of our equation when we get to dB effectiveness.

μ_o	$4\pi \times 10^{-7}$	H/m	Permeability of space
ε_o	8.854×10^{-12}	F/m	Permittivity of space
σ_{Cu}	5.96×10^7	$(\Omega\text{-m})^{-1}$	Conductivity of copper
σ_{Al}	3.78×10^7	$(\Omega\text{-m})^{-1}$	Conductivity of aluminum
Z_o	376.7	Ω	Impedance of free space

Figure 5.16: Table of useful constants

$$E = E_o e^{-x/\delta}$$

In this equation, E_o is the field strength at the surface of the shield material, x is the distance into the shield, and δ is a physical property called the *attenuation constant* or the *skin depth*.

Convert this to dB to get the effectiveness by absorption:

$$A = 20\log_{10}\left(\frac{E_o}{E_o e^{-t/\delta}}\right) = \frac{t}{\delta} 20\log_{10} e$$

The result is a very simple formula for absorption effectiveness, where t is the thickness of the shield:

$$A = 8.69\frac{t}{\delta} \quad \text{in dB}$$

The skin depth δ depends on the frequency ω in radians per second, the permeability of the material μ, and the conductivity of the material σ:

$$\delta = \sqrt{\frac{2}{\omega\mu\sigma}}$$

Skin depth can also tell us something about choosing conductors, which sounds a bit strange. Let's calculate the skin depth at 60 Hz:

$$\delta = \sqrt{\frac{2}{(2\pi)(60)\left(4\pi\times10^{-7}\right)\left(5.96\times10^7\right)}}$$
$$= 8.4 \quad \text{mm}$$

This says that the electromagnetic field is only one-third (actually, 1/e) as strong about one centimeter into the copper conductor, so the middle of the conductor carries less current than the outside does. The message: It isn't economical to use very large conductors—several smaller ones are less wasteful of copper.

5.4.2 REFLECTION BY THE SHIELD

Reflection of electromagnetic energy is not intuitive, although we are certainly familiar with mirrors. Energy reflected from a shield surface reflects in much the same way as light does from the mirror.

Light reflection depends on the difference between the indices of refraction of the air and the glass (or whatever materials we are dealing with). Electromagnetic energy reflection depends on the difference between the impedances of air and the shield. So we need these impedances.

The impedance of free space (and hence of air) is

$$Z_0 = \sqrt{\frac{\mu_0}{\varepsilon_0}} = 377 \ \Omega$$

The impedance of the shield material is

$$Z_m = \sqrt{\frac{\omega\mu}{\sigma}}$$

where μ will be μ_0 unless the shield is made of an iron-bearing or other magnetic material.

Reflection R in dB in terms of impedances is

$$R = 20\log_{10}\left(\frac{377}{4Z_m}\right) \ \text{in dB}$$

5.4.3 A AND R TOGETHER

Shield effectiveness S is the sum of absorption and reflection:

$$S = A + R$$

$$= 8.69\frac{t}{\delta} + 20\log_{10}\left(\frac{Z_0}{4Z_m}\right) \ \text{in dB}$$

where t is the shield thickness and δ is the skin depth.

5.4.4 EXAMPLE II–SHIELD THICKNESS

I want an aluminum shield that will have a shield effectiveness of 80 dB at 300 MHz. How thick should it be?

I'll try reflection R first because I've noticed it produces higher numbers unless frequencies are extremely high (well above 300 MHz):

The surface impedance of aluminum is

$$Z_m = \sqrt{\frac{(2\pi)(300 \times 10^6)(4\pi \times 10^{-7})}{3.78 \times 10^7}} = 7.92 \ \text{m}\Omega$$

That makes reflection come out to be

$$R = 20\log_{10}\left(\frac{377}{(4)(7.92\times10^{-3})}\right) = 81.5 \text{ dB}$$

Oh, my! That comes out larger than 80 dB, so I don't need to worry about absorption A.

To show how absorption works, though, let's increase the requirement to have a shield effectiveness of 100 dB at 300 MHz. That means we need some help from absorption because

$$S = 100 = A + R = A + 81.5$$

This means that absorption must provide

$$S = 18.5 = 8.69\frac{t}{\delta}$$

To find t I need the skin depth for aluminum at 300 MHz:

$$\delta = \sqrt{\frac{2}{(2\pi)(300\times10^6)(4\pi\times10^{-7})(3.78\times10^7)}}$$
$$= 4.73\times10^{-6} \text{ m}$$

Solving for t gives us a shield thickness of

$$t = 10.1 \ \mu\text{m}$$

Note that the value is in *micro*meters! We can accomplish this with a plating of aluminum on Mylar, something that is commonly done in cables (like the shield in Fig. 5.13).

5.4.5   HOLES IN THE SHIELD

OK, how about holes? I have this circuit to shield, and I've figured out I need an aluminum box to do it. But I have a couple of cables that have to enter the box, so I'll need holes. Won't the holes "let in" electromagnetic energy and mess up my shielding?

Nope! Not as long as the holes are "too small" for much of the energy to squeeze through. Well, OK, EM waves don't squeeze through things, but in a way, this isn't a bad description of what's going on. If the wavelength of the wave is too long, a hole of a given size attenuates the wave.

Let's say what I just said in terms of frequency. There is a *cutoff frequency* associated with a hole. Below this frequency (i.e., at *longer* wavelengths), holes both reflect and attenuate the wave. Above it, the energy goes through.

The cutoff frequency is

$$f_c = \frac{c/2}{L} = \frac{1.5\times10^8}{L} \text{ Hz}$$

c is the speed of light

where L is the longest dimension of the hole in meters. This says that energy whose wavelength is less than 2L goes through the hole. For example, a flashlight will shine through it.

Below the cutoff frequency, holes *reflect* electromagnetic energy:

$$R \cong 20\log_{10}\left(\frac{f_c}{f}\right) \text{ for } f < f_c$$

where f is the frequency of the incoming wave.

Holes also *attenuate*. Here are a couple of simple formulas for round and rectangular holes:

$$\text{Round: } A = 32\frac{t}{d} \text{ dB}$$

$$\text{Rectangle: } A = 27.3\frac{t}{L} \text{ dB}$$

where t is the shield thickness, d is the diameter of a round hole, and L is the long dimension of a rectangular hole.

The shielding effectiveness of a single hole is

$$S = A + R$$

N identical holes reduce the shielding effectiveness: The shielding effectiveness of a single hole is

$$S = S_{one\ hole} - 10\log_{10} N$$

The holes in the screen in your microwave's door are designed to reflect and attenuate the energy so very little gets out.

5.4.6 BUILDING SHIELDS

All of this has given us a way to find such information as how thick a shield needs to be, how big holes can be, and so on. How do we build a shield? Here are some considerations:

- Mechanically strong—which is sort of obvious!

- Low resistance, which means high conductivity such as aluminum or copper

- Carefully made joints, meaning high contact pressure or welds

- Clean joints without paint, oil, dirt, etc.

- Avoidance of dissimilar metals, which can lead to corrosion

- Doors and similar covers need contact fingers to bridge any gaps

A good example of how shielding is done is to look in the case enclosing your desktop computer. Note holes and how they are covered. Note how covers are connected by contact fingers. The shield for your computer is trying to keep electromagnetic energy *inside* the box to stay below FCC maximum EM radiation limits.

5.5   MORE EXAMPLES

Here are a couple of examples to help bring all this together.

5.5.1   EXAMPLE III–SHIELDING A CABLE

The temperature of a certain vessel is being sensed by a thermocouple that is connected to the control room by cable. The vessel temperature is about 500 °C and is to be read in the control room to a precision of ± 0.1 °C. The designers want the level of any interference on the cable to be no larger than one-tenth of the desired precision. How do we do this?

First, we need to know how much of a voltage difference one-tenth of 0.1 °C is. The thermocouple used is a Type E thermocouple, which is one of the more sensitive alloy combinations. Data from a table of voltages for this thermocouple give us the voltage we are dealing with:

$$\text{At 500 °C, } V_E = 37.005 \text{ mV}$$
$$\text{At 501 °C, } V_E = 37.086 \text{ mV}$$

Hence, the voltage difference for a precision of 0.1 °C will be

$$\Delta V = \frac{37.086 - 37.005}{10} = 8.1 \; \mu V$$

Therefore, the noise-voltage limit becomes

$$V_N = \frac{8.1}{10} = 0.81 \; \mu V$$

Now we need to know something about the cable and the environment around the cable. We learn that the cable is 40 feet long, the two wires are spaced 0.05 inches, and there are two twists per inch. Measurements and estimates tell us that the primary noise interference is from 60-Hz equipment and the magnetic flux density B is at most 2×10^{-6} webers per square meter at 60 Hz.

Remember that the voltage coupled into a loop by a B field is

$$V_m = \omega BA$$

If we ignore the twists, the voltage coupled into the cable (after converting dimensions to metric) is

$$V_m = (2\pi)(60)(2 \times 10^{-6})\left(400 \times 12\frac{2.54}{100}\right)\left(0.05\frac{2.54}{100}\right)$$
$$= 116.8 \; \mu V \; \gg 0.81 \; \mu V$$

Now let's count twists

$$N = 40 \times 12 \times 2 = 960 \text{ turns}$$

The interfering voltage is now just

$$V_m = \frac{116.8}{960} = 0.122 \ \mu V < 0.81 \ \mu V$$

and that's well under our limit. The twisted-pair cable will do the job without further shielding. However, good practice recognizes that there is probably other interference, so a shielded cable (like Fig. 5.13) will be chosen.

5.5.2 EXAMPLE IV–HOLES IN A CASE

A equipment case is shown in Fig. 5.17. It is built from aluminum 1/16" thick. Dimensions are as

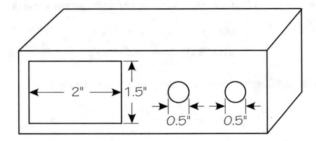

Figure 5.17: Shielded box with openings

shown. There are two holes 1/2" in diameter on the front and a rectangular hole 2 × 1-1/2" on the front.

This box is being used in an area of relatively high EM radiation. The radiation comes from a transmitter in the building that is operating on 50 MHz. The magnetic field intensity H in the area of the shielded box is near the maximum permitted by the FCC for long-term human exposure at 50 MHz, namely 0.073 A/m.

Under these conditions, what is the magnetic flux density B inside the box? Assume worst case everywhere, meaning that the EM wave arrives normal to any surface in question.

Let's start with absorption A:

$$A = 8.69 \frac{t}{\delta}$$

The skin depth for aluminum at 50 MHz is

$$\varepsilon = \sqrt{\frac{2}{(2\pi)(50 \times 10^6)(4\pi \times 10^{-7})(5.96 \times 10^7)}}$$
$$= 9.22 \times 10^{-6} \ m$$

The absorption of the aluminum case is

$$A = 8.69 \frac{\frac{1}{16} \times \frac{2.54}{100}}{9.22 \times 10^{-6}} = 1496 \text{ dB}$$

which says the field strength inside the box due to absorption is going to be negligible.

Even though the aluminum absorbs most of the electromagnetic energy, let's calculate R anyway. The surface impedance is

$$Z_m = \sqrt{\frac{(2\pi)(50 \times 10^6)(4\pi \times 10^{-7})}{5.96 \times 10^7}}$$
$$= 2.574 \times 10^{-3}$$

which makes reflection

$$R = 20 \log_{10} \frac{377}{4 \times 2.574 \times 10^{-3}} = 91 \text{ dB}$$

So even without absorption, the energy is heavily attenuated.

Each of the round holes will attenuate the field strength by

$$A = 32 \frac{1/16}{0.5} = 4.00 \text{ dB}$$

They don't cut the field by much, so maybe the round holes are a factor.

How much does reflection by a round hole reduce the field? For this I need the cutoff frequency:

$$f_c = \frac{1.5 \times 10^8}{0.5 \frac{2.54}{100}} = 11.81 \text{ GHz}$$

Our 50-MHz operating frequency is much less than the cutoff frequency. Thus, reflection is

$$R = 20 \log_{10} \frac{11.81 \times 10^9}{50 \times 10^6} = 47.5 \text{ dB}$$

So far, the round hole reduces the field by:

$$S = 4.00 + 47.5 = 51.5 \text{ dB}$$

The two round holes have a shielding effectiveness of

$$S_2 = 51.5 - 10 \log_{10} 2 = 48.5 \text{ dB}$$

Do the same thing for the rectangular hole, first finding absorption:

$$A = 27.3 \frac{1/16}{2} = 0.85 \text{ dB}$$

Better check reflection, too. The cutoff frequency is

$$f_c = \frac{1.5 \times 10^8}{2 \frac{2.54}{100}} = 2.95 \text{ GHz}$$

Again, our operating frequency is well below the cutoff frequency, so reflection at the rectangular hole is

$$R = 20 \log_{10} \frac{2.95 \times 10^9}{50 \times 10^6} = 35.4 \text{ dB}$$

The rectangular hole is worse than the round ones, which should be no surprise, since it's so much bigger:

$$S = 0.85 + 35.4 = 36.3 \text{ dB}$$

The shielding effectiveness of the rectangular hole is the poorest. Let's blame that hole for the entire field inside the box. Then the magnetic flux density inside as determined from the flux density out is

$$H_{max} = 0.073 \text{ A/m}$$
$$B_{outside} = \mu_o H_{max}$$
$$= \left(4\pi \times 10^{-7}\right)(0.073) = 9.17 \times 10^{-8} \text{ Wb/m}^2$$

S for the rectangular hole is a *de*crease in flux density, so the dB relationship is

$$S = 20 \log_{10} \frac{B_{outside}}{B_{inside}} = 36.3 \ dB$$

$$\frac{B_{outside}}{B_{inside}} = 10^{36.3/20} = 65.3$$

The flux density B inside the box, which is mostly due to the large rectangular hole, is

$$B_{inside} = \frac{B_{outside}}{65.3} = 1.40 \times 10^{-9} \text{ Wb/m}^2$$

Keep in mind that we've used a number of approximations, including some of the formulas themselves. But these combine to give us an estimate of the field inside. Once we know this, we can judge whether that is acceptable. If this isn't enough, we need to improve the shielding to reach our goal. A pretty obvious place to start with be somehow screening the large rectangular hole.

5.6   SUMMARY

Unwanted energy is coupled into circuits through capacitive, magnetic, and radiation modes, but no matter how it gets there, we don't want it. Our primary methods for keeping this energy out are to change spacing or orientation, add ground planes, or build shields.

Keeping electromagnetic radiation out makes use of both absorption and reflection. Absorption depends on the thickness of the shielding conductor. Reflection depends on the ratio of the surface impedance of the shield to that of free space. Holes, which we need for cables and ventilation, also absorb and reflect energy as long as the frequency of the energy is less than the critical frequency.

Note that the work in this chapter is not a rigorous analysis. Our goal is insight into how to keep interference out.

FORMULAS AND EQUATIONS

1. Peak voltage by magnetic-field coupling

$$|V_N| = \frac{\mu A}{2\pi r} I_o \omega = |B| A \omega$$

2. Effects of twisting conductors (max. n = 1000)

$$V_O = V_S + V_N / n$$

3. Shield effectiveness

$$S = 20 \log_{10} \frac{V_{no\ shield}}{V_{shield}}$$

4. Effectiveness is sum (in dB) of absorption and reflection

$$S = A + R \text{ in dB}$$

5. Physical constants

μ_o	$4\pi \times 10^{-7}$	H/m	Permeability of space
ε_o	8.854×10^{-12}	F/m	Permittivity of space
σ_{Cu}	5.96×10^7	$(\Omega\text{-m})^{-1}$	Conductivity of copper
σ_{Al}	3.78×10^7	$(\Omega\text{-m})^{-1}$	Conductivity of aluminum
Z_o	376.7	Ω	Impedance of free space

6. Absorption with skin depth—t thickness in meter

$$A = 8.69 \frac{t}{\delta} \text{ in dB, skin depth } \delta = \sqrt{\frac{2}{\omega \mu \sigma}}$$

7. Reflection by shield

$$R = 20 \log_{10} \left(\frac{Z_0}{4 Z_m} \right) \text{ (dB), } Z_0 = 377\ \Omega, \ Z_m = \sqrt{\frac{\omega \mu}{\sigma}}$$

8. Absorption by hole: t thick, d diam., L longest dimension in meters

$$\text{Round: } A_{hole} = 32\frac{t}{d} \text{ dB, Rectangle: } A_{hole} = 27.3\frac{t}{L} \text{ dB, for } f < \frac{1.5 \times 10^8}{L} \text{ Hz}$$

9. Reflection by hole below cutoff frequency–L longest dimention in meters

$$R_{hole} \cong 20\log_{10}\left(\frac{f_c}{f}\right) \text{ for } f < f_c, \ \ f_c = \frac{1.5 \times 10^8}{L} \text{ Hz}$$

10. Effect of N holes

$$S_{one \ hole} = A + R; \ \ S_{total} = S_{one \ hole} - 10\log_{10}N$$

Author's Biography

WILLIAM ECCLES

Bill Eccles has been Professor of Electrical and Computer Engineering at Rose-Hulman Institute of Technology since 1990 (except for one year at Oklahoma State). He retired in 1990 as Distinguished Professor Emeritus after 25 years at the University of South Carolina. He founded the Department of Computer Science at that university, and served at one time or another as head of four different departments, Computer Science, Mathematics and Computer Science, and Electrical and Computer Engineering, all at South Carolina, and Electrical and Computer Engineering at Rose-Hulman. Most of his teaching has been in circuits and in microprocessor systems. He has published *Microprocessor Systems: A 16-Bit Approach* (Addison-Wesley, 1985) and numerous monographs on circuits, systems, microprocessor programming, and digital logic design. In this Synthesis Lectures in Digital Circuits and Systems series, Bill has published several texts in this *Pragmatic* series, all to introduce electrical topics to non-electrical engineers.

Bill and his wife Trish have two children and four grandchildren. Bill is also a conductor (appropriate for an electrical engineer) on the Whitewater Valley Railroad, a tourist line in Connersville, Indiana.

He is a Registered Professional Engineer and an amateur radio operator.

Printed in the United States
by Baker & Taylor Publisher Services